Beyond The Cassandra Prophecy.

The Road to Hell.

by
Ian Gurney.

Published by IGP.
2010

ISBN 9780 9535813 3 7

The ISBN number for this book is registered world-wide with Nielsen Book Data, Nielsen Company, Nielsen Book Services Limited. (www.nielsen.com). A catalogue record for this book is available from the Library of Congress. The publishing company owning the ISBN and the publishing of this book is IGP.

"What does it matter now if men believe or not? What is to come will come, And soon you too will stand aside, To murmur in pity that my words were true."

Cassandra, in Agamemnon by Aeschylus (524 - 456 B.C.)

In Classic Greek Mythology, Cassandra, the daughter of King Priam and Queen Hecuba, was granted the power of prophecy by Apollo. However, when she refused Apollo's advances, he decreed that she would always prophecy truthfully, but would never be believed in her own land.

Chapters.

Introduction.

Now I am come to make thee understand what shall befall thy people in the latter days.

Daniel 10. Verse 14

We are out of time and on the road to hell. Mankind's tenure of planet Earth is coming to an end. We face the most significant threats to our existence that we have ever experienced and are unlikely to survive the coming apocalypse. Climate change, global warming, rising sea levels, lack of fresh water, diminishing energy resources, economic collapse, famine, disease, natural disasters and nuclear war are what await us over the next fifteen years, and there is nothing we can do to avert these catastrophes because we, mankind, are the cause of own demise.

In the following pages I will describe why our days are numbered and the terrible calamity that will explode upon the face of the Earth sooner that anyone expects. No-one will escape and mankind will be overwhelmed by the speed with which events exceed even our own worst nightmares.

Global warming and climate change now represent, according to the majority of scientists working for the United Nations Intergovernmental Panel on Climate Change (www.ipcc.ch.org) a far, far greater threat to the population of the planet than any terrorist attack. As the ice caps melt, water levels over the next ten years will rise to such an extent that the world's great coastal cities, many islands and some countries will simply disappear.

Parts of the world involved in agriculture are already finding climate change adversely affecting harvests, in many cases making agriculture impossible, leaving an ever growing population with less food to eat. Indeed, as food production falls and the world population continues to rise the price of food will escalate to such an extent that many parts of the western, developed world will begin to go hungry. Already, there are food riots taking place around the world.

Fresh water will become as precious a commodity as oil has become, with supplies around the world drying up. Energy resources will become so scarce and expensive that conflict to control these resources is inevitable.

Indeed, look around and you realise this is what is happening now in the Middle East. Diseases not seen before in the west, particularly in northern Europe and North America, will spread as global warming allows malaria, dengue fever, yellow fever and other often fatal viruses and fevers to thrive and move further north. Then there is the frightening scenario of a global pandemic, perhaps of H5 N1 Bird Flu, which could kill hundreds of millions.

As parts of the world situated around the equator reach temperatures that render huge areas of the planet uninhabitable, there will be migration on a massive scale never before seen, with perhaps three billion people seeking to move to more hospitable parts of the planet. A catastrophe of biblical proportions is awaiting us and there is no way we can stop it.

Incredibly, information about the enormous problems that face mankind has been available for almost two thousand years. The similarity between events happening in the world today and prophecies in the Bible, particularly in the books of Daniel and Revelation, point to a period of enormous catastrophe unprecedented in the history of modern humanity.

The number of the Beast, 666, has been with us

since 1973. The date of the foundation of the state of Israel in 1948, its Six Day war in 1967 and the Camp David peace accord in 1978 are all prophesied in the Bible, together with dire warnings of global famine, natural catastrophes, climate change, pollution, pandemics, earthquakes, and a nuclear war; eventually finishing with an apocalyptic event that is less than fifteen years away.

If you wish to know a little more about your future, read on.

Part One : The Earth Fights Back.

O ye hypocrites, ye can discern the face of the sky; but can ye not discern the signs of the times.

Matthew 16. verse 3

Chapter 1. Signs of the Times.

"Only two things are infinite, the universe and human stupidity, and I'm not sure about the former."

Albert Einstein (1879 - 1955)

The majority of people, particularly in the western, developed world, are stupid.

The fact that we are destroying the conditions necessary for human life on Earth to continue are staring us in the face, but we look away. We prefer instead to take our Four Wheel Drives, People Carriers and S.U.V.'s to the shopping mall for some retail therapy, or turn on the television and watch a soap opera or some "reality" TV, rather than consider the evidence placed in front of us that tells us our present lifestyles simply cannot continue.

We live in a dream world, believing that tomorrow will be similar to today and that plans we make for the future will become manifest, rather than facing the fact that our time on planet Earth is finite. We are in denial of reality and thus we sleepwalk towards disaster.

This is stupidity on a massive scale.

The United Nations Intergovernmental Panel on Climate Change, (IPCC) an organisation comprising hundreds of the foremost climate scientists in the world, issued its fourth report on the state of planet Earth to the International Climate Forum in Bali in November 2007. Since then it has upgraded its assessment of climate change and global warming. Their original predictions, they tell us, were far too conservative and in fact some of the planets ecological systems are close to or have already passed what they call the "tipping point".

This is the point where a small increase in temperature or other change in the climate could trigger a disproportionately larger, irreversible change in another eco-system. They also reveal the fact that as these systems collapse the knock-on effect to other systems increases exponentially - (like a snowball rolling down a hill, increasing in size and speed as it rolls and gathering more and more snow as it continues downwards) - leading to the possibility of sudden and dramatic changes in the worlds rainfall, ocean circulation, ice melt and sea level rise, as well as extremes of temperature and an increase in the amount and magnitude of tornadoes, hurricanes, typhoons and monsoon rains.

Professor Timothy Lenton, of the University of East Anglia in England, led a study begun in 2005 into the nature of "tipping points." Thirty six leading climate scientists drew on the expertise of a further fifty two specialists to assess the risks of "tipping points" in the Earth's climate system.

Reporting in the journal Proceedings of the National Academy of Sciences, they say :- "Climate change is likely to result in sudden and dramatic changes to some of the major geophysical elements of the Earth. Society may be lulled into a false sense of security by smooth projections of global change. Our synthesis of present knowledge suggests that a variety of tipping points could become critical within a few years."

In 2004, a report commissioned by the Pentagon,

but withheld from the public by the Bush administration, was leaked to the British newspaper The Observer (www.observer.co.uk). The report was commissioned by influential Pentagon defence adviser Andrew Marshall, who has held considerable influence on United States military thinking over the past thirty years.

The authors, Peter Schwartz, CIA consultant and former head of planning at Royal Dutch Shell Group, and Doug Randall of the California-based Global Business Network, say :- "Climate change should be elevated beyond a scientific debate to a United States national security concern." The report predicts that "abrupt climate change could bring the planet to the edge of anarchy as countries defend and secure their dwindling food, water and energy supplies."

The authors predict a very frightening scenario for the planet, saying :- "Future wars will be fought over the issue of survival rather than religion, ideology or national honour. Violent storms will smash coastal barriers rendering large parts of the Netherlands uninhabitable. Cities like The Hague are abandoned. In California the delta island levees in the

Sacramento river area are breached, disrupting the aqueduct system transporting water from north to south. By 2010 the United States and Europe will experience a third more days with peak temperatures above 90 degrees Fahrenheit. Climate becomes an "economic nuisance" as storms, droughts and hot spells create havoc for farmers. Mega-droughts affect the world's major breadbaskets, including America's Midwest, where strong winds bring soil loss. Rich areas like the United States and Europe would become "virtual fortresses" to prevent millions of migrants from entering after being forced from land drowned by sea-level rise or no longer able to grow crops."

The report concludes :- "Disruption and conflict will be endemic features of life.... Once again, warfare would define human life."

Don't forget this is the Pentagon saying this, not some liberal, left wing, environmentally friendly, tree hugging bunch of humanitarians.

The Stern Report, also presented to the International Climate Forum in Bali, was dramatic in its predictions of climate change, sea level rise and global food shortages. However, even before the report was put to the Bali conference its author, Sir Nicholas Stern said at the Royal Economic Society lecture at Manchester University, that :- "The evidence on the seriousness of the risks from inaction or delayed action is now overwhelming. We risk damages on a scale larger than the two world wars of the last century. If I was writing the report again I'd portray the risks as bigger. Things are happening far faster than anyone thought and governments must take action now."

Eco-systems are intertwined. Recently scientists have begun to understand how these fragile systems interact with each other. The oceans are the engine of the world's weather, so any major adverse changes in the oceans, particularly rises in temperature, can have similar adverse effects on the world's weather.

Some ten years ago, scientists brought to our attention one of the most serious threats to our climate ever discovered. This is a natural occurrence that changes weather patterns across the world, bringing floods where drought would normally be and drought where there is usually good rainfall. This phenomenon is called El Niño.

El Niño is a disruption of the ocean-atmosphere system in the southern tropical Pacific where, for reasons scientists still don't fully understand, the waters of the western Pacific heat up by as much as eight degrees Celsius and move eastwards towards the coast of South America. Scientists studying El Niño announced in June 1997 that they had detected the greatest climatic disturbance for 50 years, the sea temperature having risen at a faster rate than ever recorded before. Scientists say that the influence of El Niño and its sister in the northern Pacific, La Niña, on the global climate has been significant during the last few years, causing droughts, storms and high winds around the world. Indeed some say that the trade winds and ocean currents in the southern Pacific could reverse, causing havoc for global weather patterns.

Oceanographers and Meteorologists now say that the El Niño phenomenon is greatly exacerbated by climate change and global warming. They forecast that an El Niño event in the near future could be even more destructive than in 1997, causing chaos, catastrophe and famine for millions around the world.

As the oceans and the atmosphere warm, so the ice in the Polar Regions begins to melt. In the arctic, the sea ice at the North Pole is expected to almost vanish this summer. This will make no difference to sea levels as sea ice acts like the ice in a drink. As it melts the level of the drink stays the same. However, a major consequence of sea ice melt is that as the ice melts it stops reflecting sunlight back into the atmosphere. This causes the Arctic Ocean to absorb the sun's energy and precipitates a rise in the oceans temperature and a subsequent rise in atmospheric temperatures. Dr Seymour Laxon, from the Centre for Polar Observation and Modelling at University College London (www.ucl.ac.uk) reported in the 30th. October 2003 edition of Nature magazine (www.nature.com) that :- "Global warming and climate change has caused a forty percent thinning of the Arctic ice fields since the 1960's. Continued decrease in the Arctic's ice cover would also act to increase the effects of global warming in the northern hemisphere by decreasing the amount of sunlight reflected by the ice."

It is land ice, particularly glacial ice, that creates the greatest problem of rising sea levels. In the Arctic, scientists have been monitoring what they say may be the fastest moving glacier on the planet. The Kangerdlugssuaq Glacier on the east coast of Greenland has been checked using Global Positioning Satellite (GPS) equipment and was found to be flowing towards the sea at a rate of fourteen kilometres per year. Kangerdlugssuaq is seven kilometres wide, thirty kilometres long and one kilometre deep. It is also losing mass extremely fast, with its front end retreating five kilometres in 2007 alone. The glacier deposits tens of cubic kilometres of fresh water into the North Atlantic, its daily fresh water loss being equivalent to the total yearly water consumption of New York City. Dr Gordon Hamilton, of the Climate Change Institute at the University of Maine, one of the scientists studying the glacier says :- "These are very dramatic changes. The model predictions for sea level rise do not include the effects of rapid changes in ice dynamics. We're now seeing that this component might be extremely important. And what it suggests is that the predictions for both the rate and the timing for sea level rise in the next few years have been hugely underestimated."

Unlike the Arctic North Pole, which comprises of sea ice, Antarctica is an enormous land mass covered with the largest source of fresh water on the planet and, as has been mentioned, it is the melting of land ice which causes sea level rise.

Scientists estimate that the West Antarctic Ice Sheet lost about one hundred and thirty two billion tons of ice in 2006, compared with a loss of eighty three billion tons in 1996. In addition, the Antarctic peninsula lost about sixty billion tons of ice in 2006. One of the scientists researching ice loss in Antarctica, Professor Jonathan Bamber, of the University of Bristol says :- "To put these figures into perspective, four billion tons of ice is enough to provide drinking water for the whole UK population for one year. We think the glaciers of the Antarctic are moving faster to the sea. The computer models of future sea-level rise have not really taken this into account."

Sea levels are estimated to have risen by one point eight millimetres a year on average during the 20th century, but data from the past decade or so suggest that the average rise is now about three point four millimetres per year, and appears to be increasing exponentially.

Although not adding to sea level rise, the melting of the Antarctic ice shelf allows the continents land based ice sheets and glaciers to flow more quickly towards the sea. March 2008 saw the break up of the massive Wilkins ice shelf, which measures over five thousand square miles in size.

Most scientists thought that the Wilkins ice shelf would remain stable until the 2020's. However, Doctor David Vaughan, of the British Antarctic Survey (www.antarctica.ac.uk) says :- "I didn't expect to see things happen so quickly. In this case things are happening more rapidly than we thought. We didn't really understand how sensitive these ice shelves are to climate change."

According to the United Nations Intergovernmental Panel on Climate Change :- "The Earth is warming rapidly, the main cause is industrial pollution and the increase of carbon dioxide (CO_2) in the atmosphere, and the consequences for human society are likely to be catastrophic."

Scientists at the 2006 conference on Avoiding Dangerous Climate Change, held at the United Kingdom's Meteorological Office (www.met-office.co.uk) highlighted a threshold in the accumulation of carbon dioxide in the atmosphere, which should not be surpassed if a global temperature rise of two degrees Celsius above the level before the Industrial Revolution in the late 18th century was to be avoided. That threshold was four hundred parts per million of carbon dioxide in the Earth's atmosphere. Most climate scientists now believe we have passed that threshold and that four hundred and twenty four parts per million of carbon dioxide are in the Earth's atmosphere at present. The implication of this is that some of global warming's worst predicted effects, from destruction of ecosystems to increased hunger and water shortages for billions of people, cannot now be avoided

One of the scientists advising the United Kingdom government on green issues and a contributor to the findings, Professor Tom Burke, from Imperial College London, stated :-
"The passing of this threshold is of the most enormous significance. It means we have actually entered a new era - the era of dangerous climate change. We have passed the point where we can be confident of staying below the two degree Celsius rise set as the threshold for danger. What this tells us is that we have already reached the point where our children can no longer count on a safe climate."

As the temperature rises it is likely that the Greenland ice sheet will already have begun irreversible melting, threatening the world with a sea-level rise of several metres. Agricultural yields will have started to fall, not only in Africa but also in Europe, the United States and Russia, putting up to two hundred million more people at risk from hunger, and up to two point eight billion additional people at risk of water shortages for both drinking and irrigation.

A team of British scientists led by Professor Bill McGuire, head of the Benfield Greig Hazard Research Laboratory at the University College London (www.benfieldhrc.com), has found that the rise in sea levels caused by global warming could trigger off major volcanic eruptions. The team observed that ninety percent of volcanoes are close to or surrounded by sea. As water rises it erodes the lava. Eventually the mountain becomes unable to withstand the internal pressure of the molten rock and explodes. They predict the rise in volcanic activity may damage cities, have a catastrophic affect on world climate and damage air quality so badly that millions could die from respiratory ailments.

As the ice melts and sea levels rise, this presents another and unexpected problem. As Dr Seymour Laxon, from The Centre for Polar Observation and Modelling at University College London states :- "Arctic ice plays a role in the operation of the Gulf Stream, and this could be disrupted by continued thinning of the ice. It could shut down the Gulf Stream, and if that happens, Europe would be plunged into an Arctic winter within a few years."

A key factor in climate stability is the patterns of the ocean's circulation. The network of global ocean currents, of which the Gulf Stream is a part, transports not only water but also heat, with a profound effect on many regions' climate. The Gulf Stream adds about twenty per cent to the total heat from the winter Sun in northern Europe, and its shut-down would bring massive environmental upheaval.

Far from being disturbed by the effects of global warming, most people in northern Europe have welcomed this predicted climate change as having a beneficial effect on the weather, giving warmer winters and hotter summers, much like the Mediterranean climate. However, the evidence suggests that global warming could have very different effects on Europe's climate.

Currently, northern Europe and north-east America enjoy remarkably mild weather for land masses so far north. Other areas parallel to Britain for instance, such as parts of Siberia, Alaska and Canada, are inhospitable, sparsely populated and devoid of agriculture. In Churchill, Manitoba, on the same latitude as Glasgow on Scotland's west coast, the winter is long, the snow is deep, the sea freezes far and wide as the thermometer falls to minus fifty degrees Celsius. There are only two months a year without snow. When the polar bears emerge from winter hibernation, they forage among dustbins in Churchill in search of food.

The factor that keeps Glasgow, the rest of Britain and northern Europe's climate temperate is the Gulf Stream, an ocean current that brings five trillion tons of warm water from the tropics to the north Atlantic every day, around thirty million tonnes of water every second. This giant current transports so much heat to northern Europe that it plays a critical role in shaping the climate. It warms the air, and keeps winters mild. This may not last.

For over thirty years, climate researchers working for the United Nations Intergovernmental Panel on Climate Change have been analysing core samples from Greenland's polar ice caps. These tell a story of wildly fluctuating weather, with sudden and drastic changes in climate. The last 11,000 years have seen a remarkably stable period, which has enabled the growth of settlements, agriculture, and civilisation itself. But alarming new reports suggest that this period might be coming to an end.

The thick polar ice of the north Atlantic forces the warm, saline currents of the Gulf Stream deep underwater, creating an effect scientists call a "conveyor belt." As the polar water sinks, the warmer water is drawn in from the south to take its place, creating a current flowing across the Atlantic from south to north. This flow, powered by tropical Caribbean winds, adds around twenty per cent more warmth to the heat from northern Europe's winter sun.

If you pour a half glass of salty (saline) water into half a glass of fresh water, the salt water sinks because it is more dense. If the ice caps — which are composed of fresh water — start melting in sufficient quantities it could dilute the Gulf Stream, making it less saline, less dense and preventing it from sinking. It will simply stay on the surface of the Arctic Ocean and freeze. If there is no water sinking, there will be nothing to draw the warm replacement water in from the south, causing the "conveyor belt" effect to stop. According to James Hansen, Director of NASA's Goddard Institute for Space

Studies (www.giss.nasa.gov) :- "It would take no more than a quarter of 1 per cent more fresh water flowing into the North Atlantic from melting glaciers to bring the northwards flow of the Gulf Stream to a halt."

This cataclysmic event would force temperatures in northern Europe down by as much as fifteen degrees Celsius, equivalent to almost sixty degrees Fahrenheit, in a very short period of time, and according to some IPCC researchers, such a catastrophe could be imminent. The effects of a new ice age in Europe would have major consequences for the rest of the world as the entire ocean circulation patterns change and in some cases stop altogether.

The effects of such a major calamity like this are hard to comprehend. The farming industry of Europe would be completely destroyed, as the drop in temperature halts agricultural growth completely, making animal husbandry and food production virtually impossible. Europe's infrastructure, designed over centuries for a temperate climate, would collapse, forcing manufacturers, businesses and services into terminal decline, with the consequent massive rise in unemployment sending consumer spending on anything but winter clothing and food spiralling downwards. European economies, financial institutions, major businesses and services would collapse. Food would become unimaginably expensive and difficult to find. Anarchy would return to the streets of Europe.

It would appear that far from creating an idyllic Mediterranean climate, global warming could send northern Europe back into an Ice Age. As Dr Seymour Laxon says, and as core samples from the polar ice prove, this change could happen "within a few years." There would be no time for governments and people in general to react and adapt to such a dramatic change in the climate. No chance of northern Europe being able to continue a sustainable way of life. And no way anyone could do anything about it.

The Intergovernmental Panel on Climate Change's fourth report in 2007 was three hundred and sixty pages long. The summary of its findings was sixty pages long. It could have been neatly summed up in one sentence :- "Climate change and global warming will destroy mankind."

Interestingly, the idea that the Earth can rid itself of mankind was first put forward by an

English scientist who worked for many years with the National Aeronautical and Space Administration (N.A.S.A.) in the United States of America. Professor James Lovelock (www.ecolo.org/lovelock), in his 1979 book, Gaia, a New Look at Life on Earth, came up with one of the most interesting and startling theories concerning the problems that the world now appears to be facing. It is called the Gaia hypothesis. The idea behind the Gaia hypothesis is that we may have found a living organism bigger, more ancient, and more complex than anything we have discovered before. That organism is the Earth. Mankind adapts and reacts with this organism, but, as with all living organisms, if that reaction or adaptation is detrimental to the host organism, the host organism will rid itself of the problem. If proof is needed as to the validity of an argument that says the Earth can rid itself of its most damaging inhabitant, that being mankind, then look at the record of increasingly violent natural disasters and catastrophes around the planet over the last twenty years. The implications of the Gaia hypothesis are only too obvious. We are the problem.

Whatever we may consider ourselves to be,

however technologically advanced we think we are and no matter how mentally superior to all other forms of life we appear, mankind cannot fight against nature. It is the most powerful force on earth. It is now going to destroy us.

As Professor Lovelock says :- "Climate change is now past the point of no return."

His thoughts echo those of other respected climatologists and meteorologists. Sir John Houghton, co-chairman of the United Nations Intergovernmental Panel on Climate Change and former chief executive of the United Kingdom Meteorological Office, (www.met-office.gov.uk) speaking on the affects of rising sea levels caused by global warming says :- "Floods and droughts will occur in different parts of the world due to changes in the hydrological cycle. All the models we have produce the same robust result and to say nothing will happen is quite simply unreal."

And writing in the British daily newspaper the Guardian (www.guardian.co.uk) on global warming he gave this warning :-

"The Intergovernmental Panel on Climate Change has warned of one point four degrees Celsius to five point eight degrees Celsius temperature rises in the future. This already implies massive changes in climate, and yet the current worst-case scenarios emerging from the United Kingdom Meteorological Office's Hadley weather centre envisage even greater rises than this - a degree and speed of global warming the consequences of which are hard to quantify or even imagine."

Former Chief Scientific Adviser to the United Kingdom Government, Sir David King, together with a group of eminent United Kingdom scientists visited the White House in 2004 to voice their fears over global warming.

It was alleged that the White House wrote to former Prime Minister Tony Blair to complain about some of the comments attributed to Professor King, after he branded President Bush's position on the issue of global warming and climate change as indefensible, telling George Bush that :- "Climate change is a far greater threat to national security than terrorism."

Brent Blackwelder, the American Chairman of Friends of the Earth reaffirmed these opinions in a television documentary, saying :- "Global warming is affecting everything that lives and breathes upon the planet. Severe storms, hurricanes, tornadoes, huge amounts of rainfall, floods... some nations will entirely disappear. This is what is going to confront us, worse than we imagine."

Finally, the current Chief Scientific Adviser to the United Kingdom Government, Professor John Beddington, said at the beginning of 2008, that :- "Global warming and climate change will mean that demand for food will soon outstrip supply, causing an unprecedented food crisis world wide."

The response to this apocalyptic scenario from the governments of the world, particularly the developed world, has been to persuade farmers, developing countries and the population of the planet "en masse" that land now used for food production be turned, instead, to the production of "bio-ethanol" crops, such as corn, maize and rape seed oil, to power our transportation.

This, say our governments, will help reduce the emissions of potentially damaging greenhouse gases whilst still allowing us to drive our cars, fly to far off places on cheap flights and continue to ship goods worldwide by super tankers.

So, just as the world starts running out of food, our governments decide, with our approval, to reduce the amount of food being grown in the world and replace edible crops with "bio-fuels", something the United Nations recently called "a crime against humanity."

What all this means is that in ten years time we will still have our four-wheel drives parked outside our house, but no food on the table for our children to eat. Ask yourself, stupid or not?

All the information contained in this chapter is easily available in the media and on the internet and has been reported world wide on radio, television and in newspapers, so the "fool's excuse" that "I wasn't aware of what was going on" simply doesn't have any credence here.

Despite this, mankind and governments continue to look the other way and refuse to face these realities. The reasons governments will do nothing to confront these problems are simple. In order for countries to reduce CO2 emissions to levels agreed by the now outdated Kyoto Protocol, huge austerity measures would have to be introduced, with food and fuel rationing, massive tax increases and huge cuts in public spending on such things as health, welfare and pensions. Any political party putting forward such measures in their manifesto would be committing political suicide, making themselves immediately unelectable. It will only be when the public at large, you and me, start shouting at our governments to do something, that they will change policies and take the measures necessary to reduce CO2 emissions.

However, considering the general public's attitudes towards global warming and climate change, this will take some time. By then it will be too late.

> *"Science may have found a cure for most evils; but it has found no remedy for the worst of them all - apathy of human beings."*

Helen Keller (1880 - 1968)

Chapter Two. Out of Water, Out of Food.

*"What concerns me is not the way
things are, but rather the way people
think things are."*

Epictetus (55 - 135 A.D.)

Mankind is, through his reckless pollution of the planet and continuing carbon dioxide emissions, threatening the most precious resources on Earth.

In the western, developed world, if we want food we go to the supermarket. If we want water we turn on a tap. What we don't understand yet is that this will stop; and sooner than we think.

When astronomers and cosmologists search the universe for life on other planets or moons, the first thing they look for is signs of water. Without water there can be no life. The Earth is running out of water.

When rain falls it feeds the worlds rivers, many of which flow into lakes. The rain that falls on the land is absorbed through the earth's surface layers to form underground reservoirs or water tables, called "aquifers". These aquifers are accessed by drilling wells and pumping the water to the surface. The water tables under some cities in China, Latin America and South Asia are declining at a rate of over one metre per year. Water from rivers and lakes is also being diverted to meet the growing needs of agriculture and industry.

The United Nations, in its annual world population report stated that :- "Every continent has places where painful water shortages are coming. China, for example, has seven percent of the world's fresh water and twenty two percent of its population; three hundred large cities there already have serious water shortages."

The World Health Organization (WHO) estimates that over one billion people do not have access to clean water, and in developing nations up to ninety five percent of sewage and seventy percent of industrial waste were simply being dumped untreated into water courses, rivers and lakes, as well as the oceans.

In many industrial countries, chemical run-off from fertilizers and pesticides, and acid rain from air pollution require expensive and energy intensive filtration and treatment to restore an acceptable water quality. Purely technological solutions to water shortages are likely to have limited effects.

Desalinating plants that convert sea-water to fresh water, for instance, are extremely energy intensive, hugely expensive and account for less than one per cent of the water people consume. The World Bank reports that by 2020 three billion people, almost half of the world's current six and a half billion population, will live in regions facing severe water shortages.

Now, let me throw a few more figures at you.

According to the United Nations latest figures, sixty percent of available freshwater supplies are being used annually, two-thirds for agriculture. That figure is set to surge to seventy five percent by 2020 due to population growth alone, and ninety percent if consumption in the developing countries reaches the levels in the developed world.

In the 20th century, human population quadrupled, from one point six billion to six point one billion. The population has more than doubled to six point five billion in the past forty years and is projected to surge to almost nine billion within thirty years, with all the growth in developing countries whose resources are already overstretched. While global population has tripled over the past seventy years, water use has grown six-fold.

At the moment, according to the United Nations, one point four billion people on the planet live on less than one dollar a day. Nearly sixty percent of the four point four billion people in developing countries lack basic sanitation and almost a third do not have access to clean water. In the year 2000, over half a billion people lived in thirty one water-stressed or water-scarce countries. Demand for water world wide, from rivers, lakes and aquifers is increasing at a faster rate than at any time in history. At the same time, water is becoming an increasingly scarce commodity, due to pollution, irrigation, industrial use, population increase and a changing climate that is bringing storms and floods to arid areas of the world and drought to areas where rainfall would usually occur.

In 2007 Australians experienced their largest drought ever, with the Murray Darling river basin, the largest of Australia's river systems, being declared a disaster area by the government, which ordered a halt on water use for irrigation. This caused the Australian wheat harvest to fail and had a disastrous effect on south east Australia's wine growing areas.

The Murray darling basin, which is the size of France and Germany combined, produces over forty per cent of Australia's fruit, vegetables and grain. Agriculture products worth more than ten billion Australian pounds are normally exported from the region annually to Asia and the Middle East. At the same time, Australia's government scientists warned that the country, the world driest populated continent, would have increasing heat waves and predicted that droughts would occur twice as often and affect twice the usual area. Tony Burke, Australia's agriculture minister described their findings as "More like a disaster novel than a scientific report". William Cosgrove is vice president of the World Water Council, an international organization that deals with ecological problems involving water. Speaking at the Stockholm Water Symposium in 2007, he said :-

"In developing countries, irrigation today accounts for more than eighty percent of the water consumed. There is not a lake left on the planet that is not already being affected by human activities. We're killing the lakes, and that could be disaster to the human communities that depend on them. Humans are already using more than fifty percent of the world's usable freshwater resources, and ninety percent of this is in freshwater lakes. A worldwide water shortage is likely to worsen severely over the next ten years, affecting billions of people in an unprecedented global crisis."

An extreme example of this is Lake Victoria, Africa's largest lake, which has over the last two decades suffered the death of several species of fish and a dramatic increase in plant growth due to pollution from several sources, including raw sewage from surrounding towns.

"Fishermen now can't even get their boats out away from the shore to go fishing." Cosgrove said.

So, what are the consequences of diminishing water supplies to the world's population? The answer to this is simple. Starvation.

Without water for irrigation nothing will grow. Already food riots are beginning to take place around the world as the price of staples such as wheat, rice and corn double or even treble. This can only get much, much worse. All over the world farmers plant their crops in accordance with the weather, the seasons, the temperature and the rainfall. Even the seeds they plant, unless they happen to be Genetically Modified (GM) seeds, are hybrids, developed to produce the highest yield under the circumstances prevailing in the particular area where they are grown. If the circumstances in that area change, less rainfall or higher temperatures for instance, the crop will fail as the seeds are designed for a specific climate.

Even worse, farmers in the developing countries who have been persuaded to use Genetically Modified crops have found that the seeds the crops produce have been modified so that when they are replanted they will not grow. These are called "terminator seeds" and force the farmer to return to the manufacturer to buy new seeds. This is genetically planned obsolescence that seems at best immoral, and at worst, criminal behaviour. Profitable for GM seed companies such as Monsanto, disastrous for the farmers in the developing world.

So, as water for irrigation becomes more scarce, so food becomes more difficult and expensive to grow. Consider this. According to the United Nations Food and Agriculture Organization (UNFAO) it takes between two thousand and five thousand litres of water to produce one kilogram of rice, one thousand litres to produce one kilogram of wheat and five hundred litres of water to produce one kilogram of potatoes. In the developed world we think nothing of using water from the tap on a massive scale. A washing machine, for instance, uses sixty five litres of water. Flushing the toilet uses between eight and ten litres of clean, fresh water, a bath, eighty litres and watering an average garden with a hose pipe consumes five hundred and forty litres of water.

The UNFAO estimates that by 2020, three billion people will be living in forty eight countries that cannot meet the requirement of fifty litres of water per person each day for their basic human needs, let alone for growing food. As the effects of climate change and global warming increase, this figure will rise dramatically. We are already seeing the beginnings of discontent in numerous countries, some of them in the developed world, with rice, wheat and maize harvests failing, leading to an escalation in food prices and protectionism against the export of certain staple foods. India, Pakistan and Thailand banned exports of rice in April 2008. Add to this stock market speculation on the future price of these staple foods, the use of agricultural land for the growing of bio-fuels and the rising price of oil-based fertilizers and it seems the price of rice, wheat and maize will continue to rise. There is also one other problem.

As countries like China and India become wealthier, their populations are changing their diets from eating cereals, vegetables and fish to eating meat. Eating meat is a mark of affluence in many societies. This exacerbates the problem of food poverty in a very simple way. It takes eight kilograms of grain to produce one kilogram of meat. Want a hamburger anyone?

In a 2008 report, the United Nations Food and Agricultural Organisation assessed that while one hundred million tons of grain are diverted to the manufacture of bio-fuel, over seven times as much, an incredible seven hundred and sixty million tons, is used to feed animals for meat and dairy products. Enough grain in all to feed four billion people for a year. The United Nations also calculates that it takes two hundred and thirty kilograms of corn to fill a car's fifty litre petrol tank. That same amount would feed a child for a year. In many countries rainforests are being cut down to provide the land used for the production of bio-fuels. This is, according to Professor John Beddington, the United Kingdom's Chief Scientific Adviser, "profoundly stupid." He says :- "It is very hard to imagine how we can grow enough crops to produce renewable energy and, at the same time, meet the enormous increase in the demand for food."

When, in 1789, Marie Antoinette, the Queen of France, was told by an adviser that the people had no bread to eat, she responded :- "Let them eat cake."

This was the phrase that started the French Revolution. Marie and her husband Louis the Sixteenth, the King, were beheaded. The one thing that governments of any kind fear is a hungry population. They are few, the people are many and a hungry population will rise up before they starve. This is not something new, it has happened many times in the past and is starting to happen again.

It seems that mankind's indiscriminate and foolhardy use of the world's water resources has placed the population of the entire planet in a position that is completely unsustainable. We can't grow enough food to feed the population because there is not enough water on the planet to irrigate the crops. First this will cause riots. Then it will cause wars.

As the authors of the 2004 Pentagon report,

mentioned in chapter one say :- "Abrupt climate change could bring the planet to the edge of anarchy as countries defend and secure their dwindling food, water and energy supplies. Future wars will be fought over the issue of survival rather than religion, ideology or national honour."

Finally, as if things weren't bad enough already, another serious threat has recently emerged that could also have a catastrophic effect on world harvests and cause massive food shortages. In the autumn of 2006 bee keepers (apiarists) in the United States began to notice that their honey bees were disappearing from their hives. The apiarists called this Colony Collapse Disorder (CCD) when, for no apparent reason, the bees suddenly disappear from the hives, leaving just the queen, eggs and a few immature workers. Bee keepers also noticed that parasites, wildlife and other insects that would normally raid the honey and pollen from abandoned hives refused to go anywhere near them.

Since 2006 Colony Collapse Disorder has spread to half of American states, the West Coast having lost some sixty percent of its commercial bee population and the East Coast experiencing a seventy percent loss. It has also spread to

Europe, with Spain, Germany, Switzerland, Portugal, Italy and Greece all experiencing the same strange phenomena. Theories about the causes of CCD are numerous, ranging from mites, pesticides, global warming and GM crops to mobile phones. Scientific research has long shown that bee behaviour changes near power lines, now research is being carried out to see if the radiation emitted by mobile phones interferes with the bee's navigation system, preventing them from returning to the hive. The vanished bees are never found and are thought to die singly, far from the hive.

Whatever the reason, the implications of Colony Collapse Disorder are frightening. Bees make honey, but they also pollinate all our crops, fruit trees and vegetables. Without the bees to continue this pollination nothing will grow. Albert Einstein once commented that :- "If bees disappear, man would only have four years of life left."

Interestingly, a prophetic book once included in The Bible has a fascinating prediction that could well apply to the food and water crisis the world now finds itself in. The Book of Enoch is probably the most notable apocalyptic work outside the Biblical scriptures. The book was

written around 100 BC, and taken out of The
Bible by the Roman Catholic Church around 320
AD. The prediction is contained in The Book of
the Heavenly Luminaries.

And in the days of the sinners the years
shall be shortened,
And their seed shall be tardy on their
lands and fields,
And all things on earth shall alter,
And shall not appear in their time:
And the rain shall be kept back
And the heaven shall withhold it.
And in those times the fruits of the earth
shall be backward,
And shall not grow in their time,
And the fruits of the trees shall be
withheld in their time.
And the moon shall alter her order, And not
appear at her time.
And in those days the sun shall shine more
brightly than accords
with the order of light.

Enoch 2. Chapter 80. Verses 2, 3, 4 & 5

Chapter Three. Out of Oil.

"The superior person understands rightness; the inferior person understands profit."

Confucius (551 - 478 B.C.)

Governments and dictatorships, armies, aircraft, motor cars, tanks, ships, in fact pretty much everything we use comes from mankind's greatest addiction of the twentieth and twenty first century. Oil. We are running out of it.

When Rudolf Diesel (1858 - 1913) invented the combustion engine named after him and patented the device in 1893, he designed the engine to run on vegetable oil, bio-fuel as we now call it. The patent for the diesel engine was bought by John D Rockefeller, the American entrepreneur, philanthropist and founder of the Standard Oil Company (ESSO), now Exxon-Mobil, the world's largest company. When the first diesel engine was manufactured in the United States it ran, not on vegetable oil, but on diesel oil. At the time Standard Oil had a monopoly on the extraction, production and distribution of oil in the United States.

Rudolf Diesel was enraged at this and complained to Rockefeller, also writing an article publishing the fact that this was not the engine he originally invented and accusing Standard Oil of profiteering from their stranglehold on the world's petroleum resources. Six days later he died in a road accident.

The world now runs on this finite fossil fuel called oil. As stated in the last chapter, we are running out of water, a finite resource and therefore we are running out of food. Water and food are the two necessities for life on Earth to continue. Oil is a commodity that is not necessary for the survival of mankind as a species, but we have become so addicted to oil that it would seem without it life cannot go on.

We are fighting wars for control of the world's oil, because without oil there would be no transportation, no farming, no heating and lighting, no manufacturing, no plastics, no fertilizers and, interestingly, no wars. Think about this for a moment. Aircraft, tanks, battleships, helicopters, aircraft carriers and the entire armaments business could not survive without oil. You can't fly a fighter plane on solar energy, a helicopter powered by wind energy has not yet been invented. Oil is the commodity that fuels wars and this is why we will continue to fight wars for the control of oil. This is insanity of immense proportions.

Because we have not yet embraced alternative energy sources we have no choice but to chose oil, a commodity that is polluting the planet on an unprecedented scale and taking mankind to the brink of extinction.

"Civilization as we know it will come to an end sometime in this century unless we can find a way to live without fossil fuels."

That's the way David Goodstein begins his book. And that's the way he ends it. Goodstein is not an environmental extremist or a doomsayer, he is a Professor of Physics and Vice-Provost of the California Institute of Technology. (http://pr.caltech.edu)

In his 2004 book "Out of Gas: The End of the Age of Oil" (www.amazon.com) Goodstein argues forcefully that the world wide production of oil has reached its peak. This will be followed by declining availability of the fossil fuel that could plunge the world into global conflicts as nations struggle to capture their share of a declining resource.

However, Goodstein is worried that

governments are not doing enough to address the problem. "Nothing is going to happen until we have a crisis," he says, "When we have a crisis, I think attitudes will change, but by then it may be too late." That crisis, he predicts, will probably come sooner rather than later.

Most predictions concerning the end of the age of oil are based on estimates of when the supply will run out and the last drop is pulled from the last well. But that's the wrong way to look at it, Goodstein argues. Goodstein relies partly on the work of a man called Marion King Hubbert (http://hubbert.mines.edu/news/v96n1/hubbert. html) who was a historical rebel in the oil industry.

Back in the 1950s, when Hubbert was working as a geophysicist with the Shell Oil Company, he predicted that oil production in the United States would peak around 1970. He was almost laughed out of his profession, but he was proved correct. United States oil production has been declining ever since, leading to an increased reliance on imported, foreign oil.

Hubbert's formula was really quite simple. He looked at all the geological reports that were available at the time and determined how much oil nature had created underneath the United

States. Then he determined how much had been extracted. He found that half of it would be gone by 1970, and U.S. production would decline forever thereafter.

Globally, nature left about two trillion barrels beneath the ground, and the peak occurs when we reach the halfway point, Goodstein argues that :- "Since we have used close to a trillion barrels, the peak can't be more than a few years away. The worst case scenario is that after Hubbert's peak, all efforts to produce, distribute, and consume alternative fuels fast enough to fill the gap between falling supplies and rising demand will fail. Runaway inflation and worldwide depression will leave many billions of people with no alternative but to burn coal in vast quantities for warmth, cooking, and industry. The change in the greenhouse effect that results would tip Earth's climate into a new state hostile to life. End of Story."

Goodstein argues that the worldwide disruptions that will follow Hubbert's peak should serve as a global wake-up call.

"There can't be a quick resolution," Goodstein says. "It's a huge problem."

Ultimately, Goodstein argues, we must return to

the energy source used by our ancestors thousands of years ago. "There is a cheap, plentiful supply of energy available for the taking," he says, "and we won't run out of it for billions of years. It's called sunlight."

Once again, though, our governments seem incapable of investing in solar energy and alternative energy because it is vastly expensive. Meanwhile, there are fortunes to be made by multi-national corporations on the extraction, exploration, manufacture and sale of oil, and it is a fact that many of these corporations hold a massive amount of influence over the decisions made by our "democratically" elected governments.

As Dwight D Eisenhower warned in his final presidential speech in 1961 :- "In the councils of government, we must guard against the acquisition of unwarranted influence, whether sought or unsought, by the military-industrial complex. The potential for the disastrous rise of misplaced power exists and will persist. We must never let the weight of this combination endanger our liberties or democratic processes. We should take nothing for granted."

Now let me throw a few facts at you. According to the United States Energy Information

Administration (EIA) the world consumes eighty six million barrels of oil each day. America consumes almost one third of this amount, some twenty five million barrels per day, of which about seventy percent, or sixteen million barrels per day, is used by the United States military. The United States imports around fifty five percent of its oil needs at present and some commentators now estimate that by 2010 the price of crude oil could have surpassed five hundred dollars a barrel.

America represents six percent of the world's population, but contributes twenty five percent of the world's pollution through fossil fuel emissions whilst consuming thirty three percent of the world's goods. China is now almost as big a polluter as America. However, the population of the United States is only about three hundred million, compared to China's one point three billion.

The carbon footprint of the Chinese, per head of the capita, is therefore far smaller than that of Americans, but, as growth in China and the rest of the developing world continues to accelerate, this will change. The EIA estimates that China's oil consumption has grown by ten percent since 2002 and has doubled in the ten years since 1996 to six million barrels per day. India's continued growth is expected to triple its imports of oil by 2020 to five million barrels per day.

The EIA say that world demand for oil is set to increase by more than thirty percent to one hundred and fifteen million barrels a day by 2020. This figure, according to the United Nations, the United States Geological Society (USGS) and many other scientific and research institutes is completely unsustainable. There are simply not enough oil reserves in the ground to provide the world with the oil it requires. For this reason, we have gone to war in the Middle East over oil and we shall continue to fight wars in an attempt to control and dominate the finite, dwindling oil resources of the planet. This is madness when there are other forms of energy available.

However, far from exploring alternative energy, such as solar, wind and wave power, the major oil and energy companies of the world turn to increasingly expensive and polluting ways of extracting oil. In May of 2008, The Royal Dutch Shell Group withdrew it's investment in the world's largest proposed offshore wind farm, the London Array, to be situated off England's Kent coast. The wind farm would have provided power for one quarter of the population of greater London, some two million people.

The announcement of its withdrawal from the two billion pound project came days after Royal Dutch Shell revealed it had made a profit of over four billion dollars in the first three months of 2008, and on the same day that Exxon-Mobil, the world's largest company, announced three month profits of almost eleven billion dollars. Exxon-Mobil is currently in dispute with its workers in Nigeria, who want a pay rise.

Sir David King, former chief scientific adviser to the United Kingdom government, speaking about Shell's decision to the B.B.C. News (www.bbc.co.uk) warned that :-

"Energy companies need to take a long term view rather than short term profit."

Commenting on its withdrawal from the London Array, Royal Dutch Shell said wind power was vastly expensive and not as profitable as other sources of energy. The company said it intended to invest more money in the production of oil from tar sands. The extraction of oil from tar sands and oil shale is also incredibly expensive. It is also highly polluting. The amount of energy required to extract the oil is almost the equivalent of the energy it releases, the waste products from the extraction process are extremely toxic and the carbon dioxide (CO2) released enormous. It is perhaps the dirtiest way of producing oil known to man. For Royal Dutch Shell, however, it is more profitable than wind energy, end of story.

As oil becomes more expensive, more difficult to extract and increasingly dominated by the world's major powers and multi national corporations, developing countries like China and India, in order to sustain their rising prosperity and fast increasing growth, will have no alternative but to turn to coal to fire their power stations, factories, cook and keep warm, allowing even more fossil fuel emissions into the atmosphere. No thoughts of conservation, alternative energies, reduction of use or new technologies. Just a continuing use of a diminishing resource that exacerbates global warming and climate change, pollutes our atmosphere and waters and will eventually bring the world into nuclear conflict.

Stupid or what?

Chapter Four. Out of Control.

And it shall come to pass, that in all the land, saith the Lord, two parts therein shall be cut off and die; but the third shall be left therein.

Zechariah 13. verse 8

In April 2008 a conference organised by the United Kingdom's Institute for Public Policy Research (IPPR) was told by the United Nations High Commission for Refugees (UNHCR) that climate change, global warming and sea level rise could force as many as one billion people from their homes in the next fifteen years. Craig Johnstone, the deputy high commissioner for the UNHCR, said that :- "This will be a global emergency. Temperature rise across the planet will trigger mass migration on an unprecedented scale. Hundreds of millions of people will be forced to move as water shortages, crop failures and rising sea levels make them homeless."

Danny Sriskandrarajah, the head of immigration at the Institute for Public Policy Research said that :- "The displacement of millions of people will be one of the most dramatic ways in which climate change will affect humankind." Hilary Benn, the United Kingdom's Environment Secretary told the conference that :- "Climate change is the most serious long term threat to communities, with millions being forced to migrate to escape the effects of drought, flooding, food shortages and rising sea levels."

There are currently some six and a half billion people living on planet Earth. Since the beginning of the twentieth century, the population has increased to such an extent that it is now out of control. Most of this increase has been in the poorer, developing countries. Because the life expectancy in these countries is so low, couples have more children to mitigate the effects of infant mortality, famine, disease and war.

In the developed world birth rates have been falling over the last fifty years, as higher life expectancy and the lack of major natural calamities mean that infant mortality is far lower than elsewhere. It is a fact that climate change and global warming will affect poorer populations far more seriously in the first instance than those in the richer nations. Eventually, however, the richer nations will be affected just as badly as their poorer neighbours, but the affects will be compounded by the fact that, whilst poorer nations are constantly having to cope with problems of hunger, thirst, floods and other disasters, the populations of the developed, rich nations are totally unprepared for catastrophes on the scale that will occur.

As climate change, global warming and sea level rise causes millions of people to migrate from uninhabitable countries in the third world to areas where food and water is still available, what will be the effects of this huge influx of starving people to the richer nations around the world? Food shortages and lack of water are already starting to affect many of the world's richer nations, so how will they cope with the arrival of poor immigrants who put an even greater burden on a country's diminishing assets? Remember also that it is not only third world countries that will be affected by the changing weather patterns, higher temperatures and rising sea levels.

Some sixty five per cent of Holland is below sea level. The Dutch have become the greatest marine engineers in the world, designing and manufacturing the most incredible cargo ships, massive ports and huge wind powered dykes to protect the large towns and the arable countryside from flooding. However, this will not last. A rise in sea level of one meter, predicted by some scientists to happen within the next six to eight years, will inundate a huge swath of Holland, making living in these areas impossible. So what will the Dutch do when this happens?

Rising sea levels will, of course, affect many more counties than just Holland. Most of the world's major ports and important cities are situated by the ocean and are just above sea level. Think of the consequence of a one meter rise in sea level on these huge conurbations. Millions of people will be forced to move to higher places to avoid the floods, and here we're talking about London, New York, Los Angeles, Sydney, Perth, Rotterdam, The Hague, Marseilles, Tokyo, Bombay, Shanghai and many other major cities in the world. This will be a catastrophe unparalleled in human existence. What's your plan to escape from this inundation? Buy a boat?

It is now becoming obvious to the hundreds of scientists who have been involved in the studies of climate change, global warming, sea level rise, oil resources, water availability and food production, that the world is about to enter a state of affairs that the six point five billion people on the planet will not be able to deal with. Some scientists believe that up to three billion people could be affected by changes in the planets environmental and demographic ability to provide food and water to a massive population that is clearly becoming out of control. Is this the foretaste of the prophecy at the beginning of this chapter, where it concludes that two third of the world's population, that's over four billion people, will perish as the effects of changes in the planet's balances and stability make the continuing existence of mankind on Earth a real problem.

The great majority of scientists, climatologists and meteorologists now accept that global warming and climate change are caused by man-made emissions of carbon dioxide. If there was the political will worldwide to confront and challenge these problems we could possibly mitigate some of the more serious consequences that face us.

However, it seems that proposition is difficult for these people to contend with. So what exactly will our world leaders decide to do about the growing problems that the planet is bringing on the world? What will the world's politicians do to try and mitigate the terrible influx of migrants into the developed countries of the world? How will they protect their population from the emerging violence, force and anarchy that will slowly take to the streets as food, water and living become a bigger problem? What the hell are they doing at the moment to try and stabilise and reduce CO2 emissions, reduce people's addiction to diesel and petroleum and increase land availability for the growth of food? Answer. Nothing.

It is now quite clear that this world wide change in sea level, rainfall, temperature and migration will be too much for the world's resources to handle. Food shortages will spread world wide, as once abundant, productive arable areas are flooded. Water will become an even scarcer commodity across the globe and housing, even in the developed world, will become unattainable. For those who live in these areas and for those migrants who have moved there in the hope of survival, this will be a nightmare scenario of unprecedented horror.

The facts of this catastrophe are frighteningly clear. As the world gets smaller due to sea level rise, incredible temperature changes and unbelievably heavy rainfall, the world begins to realise that six and a half billion people is far too many individuals for the earth's declining resources to support. The population has become out of control and the wars that will begin as everyone attempts to stake their claim for any food, water and energy that exists will be fierce and unending. This is the beginning of mankind's travel along the road to Hell. This is also the beginning of mankind's final years on planet Earth.

> *"Population is a political problem. Once population is out of control, it requires authoritarian government, even fascism, to reduce it."*

United States Department's Office of Population Affairs.

Chapter Five. Monsters in our Midst.

"If liberty means anything at all, it means the right to tell people what they do not want to hear."

George Orwell (1903 - 1950)

The Supervolcano.

The sun and the moon shall be darkened, and the stars shall withdraw their shining.

Joel 3. Verse 15

In the first four chapters I have looked at the problems that climate change, global warming, sea level rise and many other man-made influences will have on the huge population of the planet.

However, there are some things that Nature can throw at us that we can do absolutely nothing about. Around the world there are several "natural catastrophes in waiting" that threaten to wreak havoc on the planet and to the world's population. For instance, in the heart of America lies a monster that could destroy all life on earth.

Lurking beneath the U.S.A.'s Yellowstone National Park (http://volcanoes.usgs.gov/yvo) is one of the most destructive natural phenomena in the world - a massive supervolcano. Only a handful exist in the world, but when one erupts the explosion will be heard around the globe. The sky will darken, black acid rain will fall, and the Earth will be plunged into the equivalent of a nuclear winter. It could push humanity to the brink of extinction.

Doctor Bruce Cornet is a geologist and paleobotanist who works with the United States Geological Survey (USGS) and studies Yellowstone's supervolcano. He explains the worrying situation that could cause the risk of a huge hydrothermal explosion event :- "Steam pressure builds underneath Yellowstone and the hydrothermal fluids and steam work their way upwards through fractures and vents, eventually establishing a continuous pathway for pressure to be released from the magma chamber. If that happens, the pressure in the magma chamber will drop until it reaches a critical stage when the superheated water within the magma chamber explodes. When that happens the supervolcano will explode violently, blowing out a chunk of its cap-rock and sending millions of cubic feet of ash into the atmosphere in a Pompeii-like explosion, but 100,000 times worse. As the steam venting subsides, there will be a false sense of security. People will think it was just another cyclical event and the danger is over. That will be the farthest from the truth. It will be the quiet before the storm."

Volcanoes have always been a threat to mankind. The Tambora eruption in Indonesia in 1815 killed more than 90,000 people, while the Krakatau eruption in 1883, also in Indonesia, killed 36,000. The last supervolcano to erupt was Toba in Sumatra 75,000 years ago. Evidence from ice cores extracted from the depths of the Greenland ice sheet show it changed life on Earth forever.

Thousands of cubic kilometres of ash was thrown into the atmosphere - so much that it blocked out light from the sun all over the world for some ten years. Two thousand five hundred miles away thirty five centimetres of ash coated the ground. Global temperatures plummeted by twenty one degrees Celsius. The rain would have been so poisoned by the gasses that it would have turned black and strongly acidic. Man was pushed to the edge of extinction, the population forced down to just a couple of thousand, resulting in a genetic "bottleneck" that can be observed in our DNA today. Three quarters of all plants in the northern hemisphere were killed.

For a long time scientists have known that volcanic ash can affect the global climate. The fine ash and sulphur dioxide blasted into the stratosphere reflects solar radiation back into space and stops sunlight reaching the planet. Temperatures drop dramatically and nothing grows, causing mass starvation.

Bill McGuire, professor of geohazards at the Benfield Greig Hazard Research Centre situated at the University College London (www.benfieldhrc.com) says that America's Yellowstone Park is one of the largest and most dangerous supervolcanoes in the world.

"The Yellowstone volcano can be likened to a sleeping dragon," says Professor McGuire, "whose slow breathing brings repeated swelling and sinking of the Earth's crust in northern Wyoming and southern Montana."

Professor McGuire went on to explain that :-
"Many supervolcanoes are not typical hill-

shaped structures but huge, collapsed craters called "calderas" that are filled with hot magma and are harder to detect. The Yellowstone supervolcano was detected in the Sixties when infra-red satellite photographs revealed a magma-filled caldera 85km long and 45km wide. It has been on a regular eruption cycle of 600,000 years. The last eruption was 640,000 years ago, so the next is long overdue."

Vulcanologists have been tracking the movement of magma under the park and have calculated that in parts of Yellowstone the ground has risen over seventy centimetres, almost two and a half feet, since 1923, indicating a massive swelling underneath the park.

On August 10th. 2003, the United States Denver Post (www.denverpost.com) reported that Liz Morgan, a United States Geological Survey (http://volcanoes.usgs.gov) research geologist had discovered a huge bulge underneath Yellowstone Lake that had risen 100 feet from the lake floor. The bulge is two thousand feet long and has the potential to explode at any time. Morgan was quoted as saying that :- "The inflated plain is a potential and serious hazard and possible precursor to a large hydrothermal explosion event."

"The impact of a Yellowstone eruption is terrifying to comprehend." says Professor McGuire. "Magma would be flung 50 kilometres into the atmosphere. Within a thousand kilometres virtually all life would be killed by falling ash, lava flows and the sheer explosive force of the eruption. One thousand cubic kilometres of lava would pour out of the volcano, enough to coat the whole of the United States of America with a layer 5 inches thick. The explosion would be the loudest noise heard by man for 75,000 years."

The long-term effects would be even more devastating. The thousands of cubic kilometres of sulphuric ash that would shoot into the atmosphere would block out light from the sun, making global temperatures collapse. This is called a nuclear winter. A large percentage of the world's plant life would be killed by the ash and the drop in temperature. The resulting change in the world's climate would devastate the planet, and scientists know that another eruption is due - they just don't know when.

Michael Rampino, a geologist at New York University, was quoted in a BBC Horizon documentary on Yellowstone Supervolcanoe (www.bbc.co.uk/science/horizon) as explaining

:- "It's difficult to conceive of an eruption this big. It's really not a question of if it'll go off, it's a question of when, because sooner or later one of these large super eruptions will happen."

Professor Bill McGuire says :- "There's nowhere to hide from the effects of a supervolcano. One day - perhaps tomorrow, perhaps in fifty years, perhaps in ten thousand years - it will erupt; once again wreaking devastation across the North American continent and bringing the bitter cold of a volcanic winter to Planet Earth. Mankind may become extinct."

The Mega Tsunami.

And a mighty angel took up a stone like a
great millstone, and cast it into the sea,

Revelation 18. verse 21

Scattered across the world's oceans are a handful of rare geological time-bombs which, once unleashed, create an extraordinary phenomenon, a gigantic tidal wave, called a Mega Tsunami. These are able to cross oceans and ravage countries on the other side of the world. The word Tsunami derives from the Japanese word for "harbour wave" and was a word hardly known until the Boxing Day Tsunami of 2004 that killed thousands in south-east Asia, Thailand and Sri Lanka. They are normally generated by offshore earthquakes, sub-marine landslides and undersea volcanic activity, and range from barely perceptible waves to walls of water up to 300 feet high.

The Canary Islands lie in the Atlantic Ocean off the west coast of North Africa and are well known as a winter holiday destination for millions of Europeans. What most of these holidaymakers are unaware of is that on one of the Canary Islands lies a major global catastrophe in the making, a natural disaster so big that it could flatten the Atlantic coastlines of Britain, Europe, North Africa and the entire east coast of the continent of America.

Recently scientists have realised that the next Mega Tsunami is likely to begin on one of the Canary Islands, where a wall of water will one day race across the entire Atlantic Ocean at the speed of a jet airliner to devastate the east coast of Canada, the United States, the Caribbean and Brazil.

Dr Simon Day, who works at the Benfield Greig Hazards Research Centre, University College London (www.benfieldhrc.com) says that one flank of the Cumbre Vieja volcano on the island of La Palma in the Canaries, is unstable, and could plunge into the ocean during the volcano's next eruption. During the last eruption of Cumbre Vieja the western flank of the volcano split and slid twenty metres towards the sea.

Dr. Day says :- "If the volcano collapsed in one block of almost twenty cubic kilometres of rock, weighing five hundred billion tonnes — the size of Jamaica — it would fall into water almost four miles deep and create an undersea wave two thousand feet tall. Within five minutes of the landslide, a dome of water about a mile high would form and then collapse, before the Mega Tsunami fanned out in every direction, travelling at speeds of up to five hundred miles per hour. A three hundred and thirty foot wave would strike the western Sahara in less than an hour."

Europe would be protected from the fiercest force by the position of the other Canary Islands, but the tsunami would still bring thirty three foot waves to Lisbon and La Coruña in Portugal within three hours.

After six hours it would reach Britain, where waves up to forty feet high would hit southwest England at five hundred miles per hour, travel a mile inland and obliterate almost everything in its path. Even Britain's more sheltered shores, in the North Sea and Irish Sea, will be struck by smaller but still significant swells, causing widespread flooding in major coastal cities.

"We need better models to see what the precise effects on Britain will be." Dr. Day said. However, it is likely that London could suffer severe inundation as the Thames Barrier, the flood barrier protecting London from the river's tidal surges, could not cope with such a dramatic rise in water levels.

"The Thames estuary is already subject to major tidal surges," says Dr. Day, "and the Mega Tsunami could raise water levels by as much as twenty feet, with the surge travelling up the river at some two hundred miles per hour."

Devastation along both banks of the River Thames would be huge, with many parts of the city of London and areas along both the north and south banks of the river as far as Putney Bridge and beyond experiencing severe damage. The effects on London's underground rail system are hard to imagine, but the entire network would be flooded and the consequent loss of life would be immense.

However, the destruction in the Britain would be as nothing compared to the devastation reeked on the eastern seaboard of the continent of America. Dr. Day claims that the Mega Tsunami will generate a wave that will be inconceivably catastrophic. He says :- "It will surge across the Atlantic at five hundred miles per hour in less than seven hours, engulfing the whole Canadian, United States and South American east coasts as well as the Caribbean islands with a wave almost two hundred feet high, sweeping away everything in its path up to twenty miles inland. Newfoundland and Nova Scotia would be hit first, followed by Boston and New York, then all the way down the coast to Miami, the Caribbean and Brazil. This would be followed by a second wave, smaller but travelling at the same speed, about two hundred miles an hour as it reaches land, which would cause even further massive destruction."

Millions would be killed and as Dr. Day explains :- "It's not a question of if Cumbre Vieja collapses, it's simply a question of when."

The Asteroid.

And the second angel sounded, and as it were a great mountain burning with fire was cast into the sea.

Revelation 8. verse 8

On September 29th. 2004 a three mile long asteroid made the closest approach of any asteroid or comet to Earth during the last thirty years, according to the scientists at NASA's Jet Propulsion Laboratory Near-Earth Orbiting Programme (http://echo.jpl.nasa.gov/asteroids). The asteroid came within nine hundred and sixty three thousand miles of Earth. In cosmic terms, a very near miss.

Asteroid 4179 Toutatis was named after a Celtic/Gallic god whose name is often invoked in the well known comic book series "The Adventures of Asterix," set in ancient Gaul. Toutatis is the protector of Asterix and his compatriots, who fear nothing except that someday the sky may fall on their heads. It is one of the largest known "Potentially Hazardous Asteroids" (PHA) that approaches our planet on a Near Earth Orbit (NEO). Close encounters with Venus, Earth, Mars and Jupiter constantly alter the shape of the asteroid's path as it loops through the solar system every four years. On October 31st. 2000 the asteroid passed less than 29 lunar distances from Earth. The September 29th. fly-pass came within four lunar distances of the Earth.

Toutatis has one of the strangest rotations yet observed in the solar system. Instead of the spinning about a single axis, as do the planets and the vast majority of asteroids, it "tumbles" somewhat like an American or rugby football when it bounces. At three miles long Toutatis is only half as big as the asteroid that wiped out the dinosaurs sixty five million years ago. This asteroid was seven miles wide and hit the Yucatan peninsula and formed the Gulf of Mexico. Apart from destroying the dinosaurs, it also caused the extinction of three quarters of all the living species on the planet.

Toutatis travels at a speed of about twenty miles per second and if it struck the oceans would unleash a "mega tsunami" that could reach around the entire globe, inundating millions of hectares of land, destroying coastal habitations and killing millions of people. If it hit land it would completely destroy an area the size of

Europe and would raise enough dust into the atmosphere to change the climate of the planet completely, causing a mini ice age that would freeze crops, destroy plant life and pre-empt a global famine. It would also completely destroy the ozone layer. Mankind would probably become extinct.

In 1908, an asteroid measuring about one hundred metres across, far smaller that Toutatis, exploded four miles above the Earth, releasing the energy of a thousand Hiroshima bombs over Tunguska in Siberia, flattening trees over a thirteen mile area and killing hundreds of reindeer. If the asteroid had landed in New York the city would have been completely destroyed killing millions. Tunguska thankfully impacted an unpopulated part of Siberia, but caused huge damage to the area below the explosion. Dr. Duncan Steel, an asteroid expert at the University of Salford in northern England, said :- "If the next Tunguska explosion occurred over Marble Arch in the centre of London, the whole of London out to the M25 orbital motorway would be demolished."

Toutatis is not the only asteroid that presents a threat to planet Earth. Scientists at the Jet Propulsion Laboratory Near-Earth Orbiting

Programme in Pasadena, California are currently monitoring some four thousand Near Earth Orbiting asteroids of different sizes, some of which are classified as Potentially Hazardous. The discovery of these objects over the last few years has led many astronomers to call for a concerted international effort to identify asteroids and comets whose orbits could, in the future, cross earth's path and collide with the planet. The effects of any cosmic collision with earth would be catastrophic, and scientists and astronomers have recently found evidence of past collisions that have caused enormous damage. The problem with these near earth objects is that scientists and astronomers are not aware of their closeness to earth until it's almost too late. On Friday 1st. September 2000 the 2000 QW7 asteroid, which originated in the asteroid belt between Mars and Jupiter, passed within 2.4 million miles of the earth. In cosmic terms this was a near miss. A scientist equated its closeness to earth as being comparable to a man at one end of a tennis court throwing a marble at a man at the other end of the court and missing his head by the width of a hand. Scientists say that if 2000 QW7, which is nine times larger than the Tunguska asteroid, had impacted in the Atlantic Ocean, everything within two miles would have been vaporised and the east coast of the United States and the west coast of Europe would have been swept by massive tidal waves. Molten debris would rain down for weeks after the impact, and dust particles thrown up into the atmosphere would block out the sun and have the same effects as a nuclear winter. Nothing would grow, the earth would starve. The worrying fact is that the asteroids approach was detected by Cornell University only six days before it hurtled past earth.

The Earth has been incredibly lucky since the last major asteroid strike sixty five million years ago, but most scientists agree that sooner or later we will be hit again. They know that they will be able to identify the asteroid that will collide with the planet, but admit that the world will only get a few weeks or perhaps a few days warning before it strikes. Of course, if an asteroid the size of Toutatis hit Earth, there would be nowhere to run, nowhere to hide.

Perhaps the best position to adopt in that kind of situation is to stand with your legs apart, put your head between your knees and kiss your ass goodbye. Do it quickly though.

The Earthquake.

And there was a great earthquake, such as was not since men were upon the earth, so mighty an earthquake, and so great.

Revelation 16. verse 18

There is one further natural disaster that Nature can throw at us that can have devastating effects wherever they occur and which we can do very little about. Furthermore, these events happen on a daily basis and there are many more of them than most people imagine.

Earthquakes occur along the worlds fault lines. The continents "float" on massive tectonic plates that in turn float on the magma or molten rock deep below the Earth's surface. Where the different tectonic plates meet they put enormous pressure on each other, causing rifts in the earth's crust and forming fault lines that run round the planet. That these tectonic plates are constantly moving is witnessed by the number of earthquakes that occur along the world's fault lines every day, as can be seen by visiting the United States Geological Society's website (http://earthquake.usgs.gov)

The 6.5 magnitude Iranian earthquake of December 26th. 2003, which destroyed the Silk Road city of Bam in South-Eastern Iran, is reported to have killed some 40,000 people.

The San Andreas fault runs through California for 800 miles, from the Mexican border in the south to the Oregon border in the north, and marks the boundary between two great land masses that are slowly moving in opposite directions - the Pacific tectonic plate on the west and the North American plate on the east. As the plates grind against one another, earthquakes are triggered along the fault line, which was responsible for the 1906 San Francisco earthquake.

The largest earthquake to have occurred in the United States of America was a magnitude 9.2 that struck Prince William Sound, Alaska on Good Friday, March 28th, 1964. The largest recorded earthquake in the world was a magnitude 9.5 in Chile on May 22nd, 1960.

Scientists now point to the inevitability of a severe earthquake, probably above magnitude 9 along the San Andreas fault in California in the near future. The San Andreas fault is the most studied fault line on the planet. The Pacific tectonic plate is enormous, running as it does down the coast of America, South America, across the Pacific and upwards through New Zealand, Indonesia, the Philippines, southern China, Japan and the Kuril Islands, along the so called "ring of fire". Scientists studying the San Andreas fault say that the "big one" when it occurs could be of such force that the entire Pacific plate may move causing massive earthquakes along its perimeter. Not only would California be devastated but Japan, a country notorious for its earthquakes, could quite literally be cut in half. The loss of life would be incalculable.

The other problem that an earthquake of such huge magnitude could bring is the recently developed scientific evaluation of major earthquakes and their disproportionately larger, irreversible changes on other tectonic plates, earthquake zones and volcanic systems. Some scientists studying the San Andreas fault over the last twenty years are now saying that a "big one" on California's west coast could not only affect the entire circular Pacific tectonic plate, but have serious implications for the Yellowstone National Park's enormous supervolcano, as well as posing a threat to the Cumbre Vieja volcano on the Canaries island of

Las Palma. Should such a huge earthquake wake these natural monsters from their sleep, the results to the planet and the population of the world would be catastrophic. More than a billion people would die.

If California were a country, it would be the fifth wealthiest country in the world, Japan is, after America the second wealthiest nation in the world, although China, whose east coast lies on the Pacific plate, is rapidly emerging as a major player in the world's finances and markets, together with India and the rest of south-east Asia. Should an earthquake devastate both California, Japan, south-east Asia, and parts of China, the consequences to the world's financial markets would be apocalyptic. Stock markets around the world would crash, banks and insurance companies, major investors in both California, Japan and China would become insolvent and people's investments and savings would be wiped out. In short, forget about the credit crunch, in these circumstances the world's economy would simply collapse. Scientists studying the San Andreas fault are all in accord about the "big one". It will occur, the only question is when? Enjoying living in California?

Part Two. Mankind Fights Back.

"He who joyfully marches to music in rank and file has already earned my contempt. He has been given a large brain by mistake, since for him the spinal cord would fully suffice. This disgrace to civilization should be done away with at once. Heroism at command, senseless brutality, deplorable love-of-country stance, how violently I hate all this, how despicable and ignoble war is; I would rather be torn to shreds than be a part of so base an action! It is my conviction that killing under the cloak of war is nothing but an act of murder."

Albert Einstein

Chapter 6. Chemical and Biological Weapons.

"I know not with what weapons World War III will be fought, but World War IV will be fought with sticks and stones."

Albert Einstein

When one begins to consider the scenario that

now faces us at the beginning of the twenty first century, it is obvious that something has gone very badly wrong in the world. The tragic and cataclysmic events of September 11th. 2001 in New York City have been interpreted by many as the beginning of a new, radical and frightening change in the way wars are to be fought in this brave new world. Others see it as a wake up call to the west and a moral challenge to the prosperous countries on the planet, who continue to endorse and oversee a political and financial stranglehold on the fragile economies of the third world, to rethink their global strategies.

There can be no doubt that September 11th. changed the world forever. The usual rules of engagement have disappeared, to be replaced with a theory that simply states that there are no longer any rules. Anything goes. The terrorist attacks on the World Trade Centre and The Pentagon were described by the majority of the world as an act of huge evil; and yet the immediate reaction amongst those very same people was to sanction vengeful retribution on others, mainly in Muslim countries, who where in most cases as innocent as those who perished in the Twin Towers. The perception here is one of evil being met with evil.

> *"Men regard it as their right to return evil for evil - and, If they cannot, feel they have lost their liberty."*

Aristotle. (384 - 322 B.C.)

Is this the way wars are to be fought in the future? And if so, how much evil will we be prepared to endure before there is a realisation that evil for evil's sake is not the answer? Doctor Martin Luther King Junior had this to say on the subject :- "Darkness cannot drive out darkness; only light can do that. Hate cannot drive out hate; only love can do that. Hate multiplies hate, violence multiplies violence, toughness multiplies toughness in a descending spiral of destruction. The chain reaction of evil - hate begetting hate, wars producing more wars - must be broken, or we shall be plunged into the darkness of annihilation."

Philosophers and theologians through the ages have attempted to constrain the barbarity, immorality and civilian deaths that occur in all wars. The most famous of these we owe to Saint Thomas Aquinas, the 13th-century theologian and philosopher. In his treatise "Summa Theologica" he laid out the moral principles that should shape our ethical thinking :-

"War has to be a last resort. It can be sanctioned only by a legitimate authority and can be fought only to redress an injury, with self- defence the obvious justification. Even then, a war can be fought only if there is a realistic chance of success. War's ultimate goal must be the re-establishment of peace and the peace secured afterwards must be superior to that which would have prevailed if war had not been fought. Violence used in the war must be proportionate to injury suffered. Methods of waging war must try to distinguish between combatants and non-combatants. Civilian deaths are justified only if they are the unavoidable consequences of destroying an offensive military target, not as a means to an end."

Aquinas's theory has since been honoured as much in the breach as in the observance, and as Blaise Pascal the 17th-century French mathematician, philosopher and theologian commented :- "Men never do evil so completely and cheerfully as when they do it from religious conviction."

There have been persistent attempts to regulate the conduct of war. International accords, particularly the Geneva Convention, try to tie states to the doctrine of restraint. There are philosophical objections to the "just war" doctrine, and not solely from pacifists. Many people argue that all means are potentially legitimate to minimise the length and cost of war. With such thinking, those responsible for the bombing of Dresden, the destruction of Nagasaki and Hiroshima, the obliteration of Cambodia and the destruction of the Iraqi people, seek justification.

However, such justification opens up a veritable can of worms, particularly when we consider the enormous arsenal of weapons of mass destruction now available around the world, making the problems the world faces grow ever more terrifying. How, for instance, would mankind survive an attack on the population by a chemical or biological weapon.

According to Robert Harris and Jeremy Paxman's book "A Higher Form of Killing," (www.amazon.com) the history of bio-weapons began during World War II, when the Japanese cultivated the plague-infected flea as a biological weapon. Pingfan (a bio warfare laboratory in Japan) was said to be capable of producing 500 million fleas a year.

Following the war, that technology was warmly embraced by America's bio warfare engineers, who had their Japanese counterparts flown over to the United States to share the tricks of their trade. Fort Detrick in Maryland, the long-time home of American biological warfare research, soon became the world's premier site for developing such weapons of war as the "flea bomb".

Among the potential agents studied at Camp Detrick were anthrax, glanders, brucellosis, tularemia, meliodosis, plague, typhus, psittacosis, yellow fever, encephalitis and various forms of rickettsial disease; fowl pest and rinder-pest were among the animal viruses studied; various rice, potato and cereal blights were also investigated.

In 1956 the United States army began investigating the feasibility of breeding fifty million fleas a week, presumably to spread plague. By the end of the fifties the Fort Detrick laboratories were said to contain mosquitoes infected with yellow fever, malaria and dengue (an acute viral disease also known as Breakbone Fever, for which there is no cure); fleas infected with plague; ticks contaminated with tularemia; and flies infected with cholera, anthrax and dysentery.

It would appear then that the United States has a long history of researching and developing infected insects as biological warfare agents. Just one week before the September 11 attacks, the New York Times (www.newyorktimes.com) reported that US biological weapons research was still very much alive and well, though cloaked as always in "defensive" research :-

"Over the past several years, the United States has embarked on a programme of secret research on chemical and biological weapons that, some officials say, tests the limits of the global treaty banning such weapons. The projects, which have not been previously disclosed, were begun under President Clinton and have been embraced by the Bush administration, which intends to expand them."

It would seem, perhaps, that terrorists, rogue states and "the enemies of democracy" are not the only people prepared to use these terrifying weapons of mass destruction against innocent civilians in the "War on Terror."

This leads to a question which again challenges our thoughts on the very nature of good and evil, indeed, on the nature of humankind itself, in particular those in whom we trust. Why and for what purpose is the world manufacturing Chemical and Biological weapons that have the capacity to kill every living thing on the planet? Is the continuing manufacture and deployment of these weapons a good thing, a deterrent if you will; or simply plain evil, the Devil's work?

Genetic engineers already have it within their grasp to devise a lethal bio-weapon for terrorists and rogue states, the British science publication Nature (www.nature.com) has warned, and the World Health Organisation states that :- "New advances in technology have made it possible for terrorists to kill millions of people with chemical or biological weapons. Small changes in the DNA of well-known bacteria and viruses, something referred to by scientists as "weaponizing" could turn these agents into mass killers."

The publication echoes warnings by a pair of Australian scientists, Dr Ron Jackson and Dr Ian Ramshaw, who accidentally created an astonishingly virulent strain of mouse pox, a cousin of smallpox, among laboratory mice.

They realised that if similar genetic manipulation was carried out on smallpox, an unstoppable killer could be unleashed. They published their findings in January 2001 to draw attention to the potential misuse of biotechnology and warned :- "Making subtle genetic alterations to existing pathogens to increase their virulence or durability in the environment, or to make them harder to detect or to treat with drugs, is within the limits of today's technology. With the decoding of a pathogen's entire genome now commonplace, and transgenic techniques advancing all the time, some researchers believe that the sinister potential of biology can no longer be ignored."

Bio warfare - use of germs or viruses such as anthrax or smallpox - has long been considered by military strategists. However, the risk has increased thanks to advances in knowledge about how genes work, new techniques for moving pieces of DNA around, and the relative ease with which a rogue organisation could build or hire a lab to build such a weapon.

Scientists interviewed by Nature (www.nature.com) ruled out, for the time being, the ability to build new, artificial agents from a set of component parts. "A far simpler way would be to tweak the performance of an existing bacteria to make it more resistant to antibiotics." they said.

The genetic sequences of bacteria such as tuberculosis, cholera, leprosy and the plague are already in the public domain - as is that of a food poisoning bug, Staphylococcus aureus, that is already becoming resistant to antibiotics. By identifying the genes from Staphylococcus aureus that make the bug resistant, and inserting them into the other bacteria, a scientist could make a killer for which there would be no defence.

Dr Willem Stemmer, chief scientist with Maxygen, a California pharmaceutical research firm, used one of these techniques to explore how resistance genes work. He created a strain of the common intestinal bug Escherichia coli that was 32,000 times more resistant to the antibiotic Cefotaxime than conventional strains. He destroyed the super bug in response to the American Society for Microbiology's concerns about potential misuse. Harvard University molecular biologist Professor Matthew Meselson, who has often spoken of the dangers of bio warfare, said :- "It's time for biologists to begin asking what means we have to keep the technology from being used in subverted ways."

Indeed, there is a feeling amongst many in the scientific community that a Pandora's Box has been opened that could have devastating results for mankind. The ease with which these chemical and biological weapons can be manufactured has made the discovery and monitoring of illegal laboratories extremely difficult for the international security services. What the results of a major chemical or biological attack by terrorists or a rogue state would have on a major city cannot be quantified simply because it has not, as yet, happened. However, comparisons of a kind do exist.

In the United Kingdom the rabbit-infested island of Gruinard, half a mile off the Northwest coast of Scotland is a sinister reminder of Britain's wartime experiments with anthrax.

Today the 520-acre uninhabited island, once known as Base X, is home to a small flock of about 20 sheep — the first animals to be introduced to Gruinard since it was secretly contaminated with anthrax in 1942. The animals do not appear to suffer any ill-effects from grazing on land once poisoned with anthrax spores. But when they die of old age, the sheep will not be eaten. No one is willing to eat the meat of anything that lived on Gruinard. The island was chosen by the Ministry of Defence to test the viability of anthrax as a biological weapon of mass destruction. The release of anthrax spores in 1942 killed all the sheep living on the island.

The results were so horrific that the experiments were abandoned and for the next 50 years the Ministry of Defence forbade anyone to set foot on Gruinard. It was not until the mid-1980s that scientists in protective suits were allowed to test the soil. In 1990 they set about decontaminating the island using sea water and formaldehyde.

Gruinard remains a brooding presence off the coast of Western Scotland. It was declared safe seven years ago but only a handful of visitors have made the short boat trip. Local people still call it "death island" and watch with interest as scientists continue to visit the island to test the soil. When they heard about the biological tests being carried out so close to shore, the villagers feared for their own safety.

They claimed that a strong wind could easily have blown anthrax spores from Gruinard to the mainland. Roy Macintyre, the councillor for the area, said it was after the war before the local population was told what had been going on.

"It was to some extent dragged out of the authorities. Animals washed ashore were quickly taken for examination by government veterinary scientists. That's the way things were done then. It couldn't happen now." he said.

It was not until 1999 when documents relating to the anthrax experiments on Gruinard were declassified by Britain's Public Records Office that villagers realised that the British Government had used the island for germ warfare tests.

It is quite possible that we have already sown the seeds of our own destruction. If scientists around the world, whether supported by terrorists, rogue states or national governments, are prepared to manufacture chemical or biological weapons that have massive destructive power, how long will it be before they are genetically manipulating enormously virulent diseases such as Ebola, Smallpox and Lassa Fever. Perhaps this is even now being done. There can be no doubt that much of the scientific research into killer virus's concerns attempts to find a vaccine for these terrible diseases, but in order to carry out their research accurately scientists must have access to the virus that causes the disease itself. We are constantly being re-assured, both by scientists and governments, that work on these deadly and pestilential organisms is completely safe and securely controlled, but there are many instances where security at these so-called safe establishments has been badly breached.

On February 20th. 2001 Foot and Mouth Disease (FMD) was discovered in the United Kingdom. Six months later the United Kingdom government had ordered the killing of six million cattle and sheep and the livelihood of hundreds of farmers had been destroyed. The foot and mouth outbreak could have been started deliberately by someone who stole a test-tube of the virus from a laboratory.

The British newspaper The Sunday Express said that in December 2000, two months before the crisis began, a container of foot and mouth virus went missing from a laboratory at Porton Down in Wiltshire, which houses the UK government's Secret Weapons Research Establishment. The disappearance was discovered during a routine audit of the sensitive unit, which also houses smallpox, TB, anthrax and Ebola. The newspaper said there were rumours the missing test-tube, which contained the Asian Type O virus, the same rare FMD virus found in United

Kingdom sheep and cattle, could have been taken by animal rights activists. The paper went on to quote a "senior military source close to Porton Down" as saying :- "A phial appears to have gone missing from one of the labs following a routine audit last year. Ministry officials were informed immediately and an investigation was launched initially by Special Branch and then by MI5, who are interested in the activities of animal rights protesters."

Whatever the truth about how foot and mouth disease started in Britain, there can be no doubt that the epidemic has cost the country dear, not only in terms of the livestock destroyed, but in the damage that has been done to the entire farming community in Britain. If indeed it was an attack on Britain's farmers by animal rights activists it exposes the ease with which chemical and biological weapons can be used by terrorists against a country in order to destroy its agricultural economy and indigenous food supply. Since September 11th. 2001 we have started to understand that there are many new, frightening and terrible ways in which wars can be fought and perhaps the most terrifying of these is the use of biological weapons.

The nastiest of these biological agents that have

been or are being "weaponized" consists of 19 bacteria, 43 viruses, 14 toxins and 4 rickettsiae (a large parasite found in vertebrates). Here's a list of some that terrorists are most likely to use. All of them may be subject to bio-engineering, making treatment impossible. And, of course, there are likely to be others under development.

Anthrax The most popular because it is the least difficult to spread. It can get into the body through scratches and cuts or through inhalation. After one to six days it causes fever, fatigue and a cough. This briefly clears, to be followed by severe chest problems. Inhaled anthrax has a mortality rate of almost 100%. Antibiotics can work if it is caught early enough.

Ebola and Marburg Among the most horrible ways to die, these are both viruses from sub-Saharan Africa. The effects of these two viruses are similar. Ebola attacks every tissue in the body except skeletal muscle and bone. A fever appears 10 days after infection. Haemorrhages develop under the skin, which then tears, and then bleeding occurs from all orifices. Skin, including that on the tongue, can peel off and the eyes fill with blood. There is no vaccine and no cure. Ebola and Marburg are particularly well disseminated through the air.

Smallpox The last smallpox epidemic was in Somalia in 1977. It has now been eradicated except in the bio-warfare laboratories. Twelve days after infection, the symptoms are fever, aching and vomiting, after which a rash appears that can cover the entire body and turns into painful pustules. In epidemics that occurred naturally, mortality rates were between 30-40%. Again, it survives aerosol dissemination.

Plague The incubation period of bubonic plague is between 2 and 10 days. A high fever and tenderness occurs, followed by skin lesions, pustules, and dark patches on the skin. With pneumonic plague, the fever is followed by progressive respiratory failure. Untreated bubonic plague has a mortality rate of 50%, the pneumonic strain nearly 100%. Antibiotics can work, although the Russians have developed a resistant form.

Botulinum toxin Among the most toxic substances known, this can come from poor treatment of food and has been extensively developed. Symptoms, including weakness, difficulty in swallowing and breathing problems appear after a few hours, or two days at most. Muscular and respiratory failure, then paralysis, lead to death. Effective treatment is unknown.

Ricin Derived from the beans of the castor plant, it can kill within three days when inhaled. The Bulgarian dissident Georgi Markov was killed with ricin in London in 1978. Sufferers grow weak, develop a fever, vomit blood and then die.

So do we now face a bio-apocalypse or an era of bio-terrorism? Quite possibly both, especially since scientists and researchers have realised that bio-weapons are about to get very clever. The knowledge flowing from the exploding science of genetics offers the possibility of, for example, chimeric bugs - artificial organisms made up of two natural ones. The Russians may have already spliced the Smallpox and Ebola viruses and "weaponized" them, a combination so pointlessly hellish that it can only have been done as a kind of vile joke. Or we could make utterly new organisms, resistant to any antidote. We could make diseases so genetically refined that they would only target certain racial groups. We could destroy crops and animals, or, in a more humane mood, we might simply produce drugs that debilitate armies or populations by making them depressed or submissive. Genetic information may, one day, be good news for the good doctors, but at the moment it appears to be great news for the bad doctors.

The ability to destroy life from the face of the planet, unthinkable though that may be, is now in the hands of doctors and scientists who appear to justify their work by claiming it is for defensive purposes only; and that the use of these weapons would be unthinkable. But there are those, politicians, doctors and scientists as well as tyrants and terrorists, who are prepared to think the unthinkable, and worse, to commit the unthinkable. In the 1940's mankind invented one means of destroying life on the planet, when scientists split the atom and manufactured nuclear weapons. At the time Albert Einstein, whose research and theories had in many ways contributed to the manufacture of the Atom bomb, concluded :- "The unleashed power of the atom has changed everything save our modes of thinking, and we thus drift towards unparalleled catastrophes."

Now we have embarked upon a scientific journey into the darkest abyss of human wickedness, by seeking to manufacture and utilise chemical and biological weapons that, once released into the world, will eliminate not just humankind, but every living thing on the planet. And the problem with chemical and biological weapons is this :- We can't see them, we can't smell them, we can't hear them, we can't touch them and, until it's too late, we can't even feel their effects. We will only be aware of these terrifying weapons as they kill us, slowly and painfully.

Why then are we doing this? What is it in the genetic disposition of man that spurs him onwards towards his own self destruction? Surely no human on Earth would wish that their children be subjected to the horrors of chemical and biological warfare? And if this is true, for what purpose do we continue in our efforts to destroy ourselves by manufacturing these weapons? Is it the wish of all mankind that we should inevitably become extinct? Is humanity too much in love with hell ever to miss an opportunity of going there?

> *"Mankind must put an end to war, or war will put an end to mankind. This is one of the many lessons that mankind seems unable to learn."*

John Fitzgerald Kennedy, former President of the United States.

Chapter Seven. HIV-AIDS and the Flu Pandemic.

For these be the days of vengeance, that all things which are written may be fulfilled.

Luke 21. Verse 22

So, now we have discussed the threats that climate change, global warming, sea level rise, water scarcity, food shortages, lack of oil and chemical and biological weapons present to the population of the world, let's take a little look at one plague that is already killing millions of people around the world and a coming pandemic that could claim the lives of a billion people.

As we move into the twenty first century, a lethal killer virus is already on the march around the world. In June 2001 the United Nations marked the 20th anniversary of the first reported cases of AIDS with a warning that the epidemic, despite having claimed over 22 million lives already, was just in its early stages. The United Nation's joint program on HIV-AIDS, UNAIDS, said the disease had emerged as the most devastating epidemic ever and that the world had to act now to turn back the tide.

"AIDS has become the most devastating epidemic in human history. On a global scale we are only at the beginning." UNAIDS executive director Peter Piot told Reuters. "Within the first 20 years, 58 million people have become infected and more than 22 million have been killed by HIV-AIDS."

Though more than 36 million people are now living with HIV-AIDS, most of them in Africa, the disease in many parts of the world is still in its early stages.

"HIV is characterised by a relatively long gap between infection and major illness." Piot told a meeting of health workers. "Its natural dynamic is to show up first among those at heightened risk while at the same time it gradually moves across the whole of the sexually active population."

The number of AIDS sufferers in South Africa, the country with more people living with the disease than any other is four point seven million, or one in nine of the population and is expected to reach as high as seven million, whilst India and China are expected to surpass South Africa as the disease spreads through their populations, according to health experts.

UNAIDS issued its alarming forecast 20 years to the day after the first official report of what became known as Acquired Immune Deficiency Syndrome or AIDS was made. The disease had reached every continent by 1985, but today Africa is the epicentre, with more than 25 million sufferers. Of more than 10.4 million AIDS orphans worldwide, more than 90 percent live in sub-Saharan Africa, UNAIDS said.

The origins of HIV, the virus that causes AIDS, remain a medical mystery. It was, at first, widely accepted that the virus was first transmitted from chimpanzees to humans, but no one was certain how, or when, this happened. Dr. Beatrice H. Hahn, the scientist who first reported evidence that the HIV epidemic got its start in west-central African chimpanzees has found that the monkey counterpart to HIV does occur in chimps in the wild, but perhaps only rarely. Dr. Hahn and her colleagues had previously found that the west-central African chimpanzee subspecies Pan troglodytes troglodytes harbour strains of Simian Immunodeficiency Virus (SIV) that are closely related genetically to HIV.

Researchers had long suspected that HIV arose

as a result of a viral "cross-species jump" from primates to humans. The theory is that, through contact with chimpanzee blood, possibly through hunting them and eating the meat, humans were exposed to SIV. Some scientists speculate the HIV epidemic took off in Africa due to modern-day cultural changes such as increased population movement and breakdowns in traditional lifestyles.

In looking at urine and faecal samples from 58 wild chimpanzees in three African nations, Dr. Hahn's team found that SIV does occur in chimpanzees in the wild. But they found it in only one chimp, according to the report published in the January 18th. 2002 issue of Science. (www.science.com)

In October 2008 another American team of scientists studying HIV-AIDS said that they thought that the virus might have jumped from chimpanzee to mankind in Africa as long as a hundred years ago, but did not become wide spread and a pandemic until cities and towns began to be built in Africa, enticing the population to move to more populated areas.

Though these two ideas have been well publicised all over the world, there still appears to be a lack of decisive evidence to back up these suppositions and theories.

Indeed, recently the entire theory and hypothesis of the arrival and devastation of HIV-AIDS has been challenged by an American lawyer, who believes he has discovered the most terrifying inference about HIV-AIDS.

Research conducted by Boyd E Graves JD, who has spent a great deal of time trying find out the truth about HIV-AIDS, has led him to petition in the United States Supreme Court, for the United States to make a global apology for inventing the HIV-AIDS virus. His action is being fought in the US. Supreme Court, Case No. 00-9587. (www.boydgraves.com)

Graves claims it is now becoming increasingly clear that this virus was not, as scientists at first thought, passed from chimpanzees to mankind, but was probably knowingly developed by doctors and scientists, all of whom had signed the Hippocratic Oath. The other frightening conclusion of his research is that these were not doctors and scientists working for a rogue state, terrorist organisation or supposedly unfriendly nation, but scientists working for the United States government.

Graves claims that in April 1984, Dr. Robert Gallo filed a United States patent application for his invention, the HIV-AIDS Virus. The Patent number for the invention is 4647773 and details can be found at the US Patent Office website. (www.uspto.gov)

According to Graves, the scientific evidence is compelling; the AIDS Virus was manufactured as a designer bi-product of the US Special Virus programme. The Special Virus programme was a federal virus development programme that persisted in the United States from 1962 until 1978. Dr. Robert Gallo's 1971 Special Virus paper, "Reverse Transcriptase of Type-C virus Particles of Human Origin," reveals the United States was seeking a "virus particle" that would negatively impact the defence mechanisms of the immune system. The programme sought to modify the genome of the virus particle in which to splice in an animal "wasting disease" called "Visna". Graves says Dr. Gallo's 1971 Special Virus paper is identical to his 1984 announcement of AIDS, and in 2001 Dr. Gallo conceded his role as "Project Officer" for the federal virus development programme, the Special Virus.

According to Graves, and agreed with by American doctor Alan Cantwell M.D. the US Special Virus was added as a "compliment" to

Hepatitis B vaccine inoculations given to black people in Africa and homosexuals in Manhattan and San Francisco during the late 1970's and early 1980's. Shortly thereafter the world was overwhelmed with mass infections of a human retrovirus that differed from any known human disease, it was highly contagious and it could kill. The vaccine was made by the pharmaceutical company Merk.

Boyd Graves now believes that HIV-AIDS is probably an evolutionary, laboratory development of the peculiar Visna Virus, first detected in Icelandic sheep, and recently other scientists have confirmed the possibility of the laboratory genesis of AIDS.

So, it appears that HIV-AIDS may be a virus that has been manufactured and utilised by doctors and scientists with the knowledge and co-operation of the government of the United States of America. If this theory were to be proved true, one has to consider whether or not the invention and use of the HIV-AIDS virus by the United States was, according to all the laws governing the manufacture, development and deployment of biological and chemical weapons, the first act of mass bio-terrorism this world has witnessed.

Although HIV-AIDS is considered to be the most devastating epidemic in human history, there is, on the horizon, an epidemic that could be far worse and cause an enormous population decrease around the world. It would become the most frightening pandemic to affect the human population since the flu pandemic of 1918.

In a study appearing in the journal Science, (www.science.com) scientists at the Australian National University in Canberra said the genetic union of pig and human influenza viruses triggered the most deadly disease outbreak in human history, the 1918 "Spanish" flu pandemic. A key gene in the virus responsible for the 1918 pandemic was a hybrid created by the joining together of genetic sequences of pig and human influenza viruses. The pandemic, whose outbreak came just as World War One was drawing to a close, killed more than 40 million people as it spread around the globe in 1918 and 1919.

Now think about this. In the last fifty years travel around the world has become an easy, simple and cheap way of moving from one country to another. Hundreds of thousands of people fly around the world each day and international tourist journeys are now reaching eight hundred million people each year. This enormous movement of individuals would make the spreading of a flu virus impossible to stop. Already in south-east Asia the H5N1 bird flu virus has been responsible for many deaths.

In August 2008, the British House of Lords Intergovernmental Organisations Committee produced a report that stated the worries that many of Britain's politicians have about the possibility of H5N1 becoming a world wide pandemic. Ministers of the British government, who gave information to the Committee, said :- "We have been warned that an influenza pandemic is overdue and that when - rather than if - it comes, the effects could be devastating, particularly if the strain of the virus should be of the H5N1 variety that has been seen in south-east Asia in recent years."

The chairman of the Committee, Lord Soley, warned that :- "We have a pandemic twice in every century. If a pandemic occurred in a country with a developed health care system you would stop and stop it before it went round the world. You cannot have that confidence about the developing world."

Meanwhile Norman Lamb, Britain's Liberal Democrat health spokesman, said that :- "The potential for loss of life from a pandemic is massive, enormous and yet we stare a disaster in the face and we see a chaotic, uncoordinated and incoherent international response to it."

The report concluded that a flu pandemic could kill more than fifty million people worldwide, although this seems a slightly diminutive figure, and that such an outbreak would leave up to seventy five thousand people dead in Britain and cause "massive" disruption. This disruption is something that most politicians are frightened to address, and when you consider the fears that most people would have as a pandemic spread around the world, you can't blame them. That plus the fact that in August 2008 doctors said that Tamiflu, the already available protective anti viral immunisation against H5N1 bird flu, is not proving affective to people with the virus. This sounds like trouble.

Ask yourself this question. If an H5N1 bird flu pandemic suddenly spread around the world, how safe would you consider visits to the supermarket, the bank, the local pub or bar, your children's school, university, your workplace, the cinema, the hospital, your next door neighbour and many more people and places that you visit and take for granted every day of your life?

What of the people that work in these places. Would they want to be dealing with people face to face who might be passing on the N5N1 bird flu pandemic to them while doing their teaching, serving or business? What about government members, police, defence forces, hospital staff, airline and airport staff and airline passengers? A guy at the front of the plane with H5N1 bird flu sneezes on a three hour flight and an hour later his virus is being re-circulated through the planes air system to all the passengers on board. Seriously, would you consider flying?

It seems that the British government's warning of a "massive" disruption could be right on the mark. This would not be something that just affected the United Kingdom, but every country in the world; and if there is a bird flu pandemic in any country on the planet, then that country would close. End of story.

Part Three. The Prophecies.

"If you do not believe in God, and he exists, you have a problem. If you believe in God and he does not exist, no problem."

Blaise Pascal.
17th-century French philosopher and theologian.

Chapter Eight. Science or Religion.

"Science without religion is lame, religion without science is blind."

Albert Einstein.

So, we have discussed the environment, food and population problems, nasty bio-weapons, a killer plague and coming pandemics. Now comes the difficult part of this book, the Bible. When I start talking about the Bible and the biblical prophecies most people's first question to me is "Are you religious and do you believe in God?" This is, for me, a difficult question. Let me explain why. Since I was a child I was instructed in the Christian religion at Sunday school, but like most young children, I was about five at the time, the outside world where my friends played was far more attractive. So it was, that I left my Sunday school teaching and joined my friends outside. I don't know what it was that decided me to do this, but I like to think it was simply my young age.

It wasn't until much later, in my late teens and twenties, that I began to question the reasons why we were all here, what life is all about and if God exists. Since that time I have been interested enough in the world's religions to analyse the messages that each one gives to the world and consider the true nature of God. Let me also say that in studying the world's major religions, from Islam, Judaism, Hinduism, Buddhism and Sikhism to Christianity, I have found each one truly wonderful and amazing. However, not surprisingly, I am still a poor, confused individual, particularly about the interpretation of God.

Nevertheless, I have to say that, having studied religions as well as the works of many prominent scientists, cosmologists and astronomers over the years, I now definitely believe God exists. My only problem is that I have no idea who or what God is. Even though religion is firm in its belief of God, it has been the remarkable findings of scientists and astronomers and the incredible facts they provide about the wonders of the universe that have opened my mind to the fact that God exists. Just don't ask me to describe God, for that is impossible for me to do. Yes, I know I'm stupid, but then I join the ranks of the religious teachers, scientists and astronomers who also are unsure of the true nature of the creator. Maybe this is the reason why I enjoy reading the Tao Te Ching, the sixth century BC work by Lao Tsu, a Chinese contemporary of Confucius, which is concerned with the more spiritual and mystical level of being and living.

Science and religion, although at different ends of the theological spectrum, are starting to agree about the existence of God and the profound and often contentious ideas God has on our thinking, ideology, spirituality and understanding of the universe. For instance, there is an argument between science and religion regarding the true nature of good and evil. In fact many scientists argue amongst themselves over the notion that either good and evil do not exist, exist side by side, are separate and opposite to each other or, as sociologists would have it, are states of being that can be a product of social upbringing, moral education or genetic inheritance.

Religion, of course, is more defining on the subject, treating them as separate choices open to each individual, depending on their moral principles. Good is God, evil is the Devil. We should try our best to be Good in our lives and we should avoid those things which are Evil. Though this is a wholly over simplistic way at looking at these two aberrations, it is probably the one that is most easily accepted by mankind. Nevertheless, over the last twenty years there can be no doubt that people's views of good and evil have changed, as religion, particularly Christianity, has played less and less of a part in our lifestyles. This was accepted in September 2001 by Cardinal Cormac Murphy-O'Connor, the Archbishop of Westminster and leader of the 4.1 million Roman Catholics in England and Wales when he said :- "Christianity, as a backdrop to people's lives and moral decisions has now almost been vanquished. People are seeking transient happiness in alcohol, drugs, pornography and recreational sex. There is indifference to Christian values and to the

Church among many young people and, indeed, not only the young. You see a demoralised society, one where the only good is what I want, the only rights are my own, and the only life with any meaning or value is the life I want for myself. When we live in a culture which says "What I have got is what I am", we are in big trouble. There are many today who think that to believe in God is to limit one's freedom."

Cardinal Murphy-O'Connor was, of course, talking about the situation in the United Kingdom. In many other countries, America for instance, there is still a huge majority of people who believe in God and follow their religion with great conviction. Whether or not you believe this to be a true reflection of Christianity today, there can be no doubt that God and religion have increasingly been pushed to one side in favour of market forces, financial gain, material wealth and self importance. Within this context the idea of good and evil cease to exist and are replaced by a set of ideals that seem to imply that greed is good and in a strange juxtaposition of religious belief the new mantra for a new age seems to be "Do unto others before they do unto you." It is worth, therefore, looking at the effects this attitude is having on a world that increasingly is becoming divided into two separate groups. Those that have and those that do not. And if this position is to remain what are the moral implications of a world system that leaves eighty per cent of the population poor, uneducated and hungry. As Joseph Conrad, the author born in 1857 once noted :- "The belief in a supernatural source of evil is not necessary; men alone are quite capable of every wickedness."

Perhaps the most fundamental question mankind has sought to answer since the beginning of time is quite simply what are we here for? What is the point? We live out our three score and twenty years on this planet and then what? Is that all there is? If there is no God, no resurrection, no life after death, then why are we here? Are we just part of some enormous, random, cosmic accident, a freak of nature, or are we part of something much larger, a master plan perhaps, an experiment designed by unknown forces for reasons too huge for our feeble imaginations to contemplate?

Science has yet to find the answers to these profound questions, and yet deep within our psyche a feeling exists inside every human being that there must be more to life than this, and that if science is unable to provide us with the answers maybe there is more to religion than just mysticism, dogma and faith. Perhaps the questions that science poses can only be answered by the contemplation of the spiritual and the mystical that appears to be a part of us all.

As Albert Einstein said :- "The most beautiful emotion we can experience is the mystical. It is the power of all true art and science. He to whom that emotion is a stranger, who can no longer wonder and stand rapt in awe, is as good as dead."

It seems that science, far from telling us that God is dead, is saying that the universe is so complex, so enormous, so unimaginable and yet so cosmologically constant, that we can only start to comprehend its mysteries and wonders by believing in a divine creator, a benevolent force, a universal oneness that has shaped everything we know and see and understand since the beginning.

Science seems to be accepting the fact that there are questions about ourselves and the universe that cannot be explained in pure scientific terms. Quantum physics and cosmology have tried to discover the "Theory of Everything" but have so far failed to prove that a solution is even possible. Cosmologists and astronomers now point to the fact that before the "Big Bang" there was nothing, but this "nothing" had physical properties constant with the energy released during and after the "Big Bang." What they can't explain is what that cosmologically constant "nothing" was.

Was it some supernatural force, an omnipresent divinity still pervading every part of the universe, God maybe? Or was it what scientists call "Dark Matter," perhaps the most challenging astronomical discovery of recent times. We can't see it, we can't analyse it, we can't explain it, but through spectrum analysis of the background noise in the universe, scientists tell us it exists everywhere and makes up ninety percent of the matter present in the universe. However hard science tries to offer us the explanation to everything, so far it has failed. Religion too, though having had much more time to reach its conclusions, has also failed in its quest to answer this question.

No matter what you believe or do not believe though, it appears that there are some very interesting prophecies in The Bible that until a few years ago seemed to make no sense at all.

Slowly, however, people are beginning to realise that as we move into the twenty first century, some of these Biblical prophecies have a resonance with events that are taking place now. The last seven plagues described in The Bible are already with us. 666, the "mark of the beast" has been with us since 1973 and in the books of Daniel and Revelation we are given a series of four important dates, three of which have already occurred, that culminate in Judgement Day.

Whether we have any faith or belief in these prophecies is up to each individual, but it is strange how the prophecies in The Bible only seem to have been able to be properly understood as we have moved from the twentieth century into the twenty first century. Perhaps they were written in such a way that only at the time of "the latter days" would they become comprehensible. Who knows? What is most interesting is that these prophecies can be described in the most realistic way because they are happening right now, right here on earth and right in front of our faces.

Chapter Nine. 666. The Mark of the Beast.

Here is wisdom. Let him that hath understanding count the number of the beast; for it is the number of a man; and his number is six hundred threescore and six.

Revelation 13. verse 18

OK, I know the next three chapters of this book may be a little difficult for many of you, but please, bare with me, there interesting. First, a little information about the Revelation of Saint John the Divine. Revelation, the final book of the Bible, was written in the first century AD, not long after Jerusalem was attacked and destroyed by the Romans under Titus Andronicus, son of Emperor Vespasius in 70 AD.

It was written by St. John the Divine after his banishment from Ephesus, now in modern day Turkey, for preaching the gospel of Jesus Christ.

I John, who also am your brother, and companion in tribulation, and in the

Kingdom and patience of Jesus Christ, was in the isle that is called Patmos, for the word of God and the testimony of Jesus Christ.

Revelation 1. verse 8

Patmos is in the Greek Dodecanese islands, just off the coast of western Turkey. St. John received the Revelation whilst sheltering in a cave now known as the Sacred Cave. Visitors to Patmos, designated a Holy island under Greek law in 1988, will find the Sacred Cave surrounded by a small, beautiful 11th century church, halfway up the winding road between the port or Skala and the hilltop village or Chora. Inside the Sacred Cave, adorned with religious artefacts, there are two marks, one on the floor of the cave the other on the cave wall. Both these marks are surrounded by a silver halo marking the place where St John lay his head and where the archangel Gabriel stood whilst St John received the vision of the Revelation. The small church is indeed beautiful.

The Church of St John the Divine dominates the hilltop town or Chora of Patmos and is surrounded by a medieval fortress. Inside its treasury are some of the most beautiful and striking religious artefacts in Europe or Asia, including a superb library of the scriptures and a wonderful sixth century manuscript of Saint Mark's gospel. Everywhere you look are murals, mosaics and pictures of Ioannes Theologus - St John the Divine.

The prophecy at the beginning of this chapter is perhaps one of the most misunderstood and misinterpreted verses in the Bible. Most scholars, theologians, academics and commentators have attributed the Beast and the number 666 to one individual, the Antichrist. Indeed, there have been many films over the last thirty odd years depicting this idea. However, recently some Christians in America have claimed that the final verse of Revelation chapter 13 has another meaning.

Let him that hath understanding count the number of the beast;
Revelation 13. verse 18

They say that here, Revelation is telling us not only what the number is, but asking us to count the number of the beast. If you take the English alphabet and give each letter an increasing number of 6, with A = 6, B = 12, C = 18 etc. then count up all the values of each letter found in the word COMPUTER, it adds up to 666. I'm writing this on a computer.

Now I'm not saying that these interpretations are wrong, but there is a distinctive product invented in 1973 that already holds all the information that the Bible contends will be held by "The Mark of the Beast". This product was designed to be divided into two halves, with brackets on either side of the product and a dividing bracket in the middle. These three brackets represent the number six, so the product is designed to be based on the three numbers, 666. Don't believe me? Read on.

The product we're describing here was invented by a Mister George J. Laurer. In 1971 he was an employee at one of the United States principal companies, IBM. The company assigned him the task of "Designing the best code and symbol suitable for the grocery industry". What George Laurer invented was the Universal Product Code, or Bar Code. In nineteen seventy three Mister Laurer's UPC Bar Code entered the world of the consumer society. Nowadays we see the Bar Code on everything we buy, but most people are not aware that the Bar Code operates on a 666 system. The best way to check all this out is simply to go to your kitchen cupboard or refrigerator and take out several items and check the Bar Codes. Find a Bar Code with the number 6 under one of the sets of lines and study the lines that represent 6. Notice that the Bar Code is divided into two halves. Now look at the two brackets dropping below the lines at each end of the Bar Code and the bracket below the lines in the middle of the Bar Code. Compare these lines with the lines you've just studied. These three lines all represent the number 6. Confirmation of this is given by the author Bob Fraley in his book The Last Days in America (www.amazon.com) published as long ago as nineteen eighty four. On page 225 of the book Bob says :- "The interpretation of the Universal Product Code marks is most revealing in that the three numbers 666 are the key working numbers for every designed Universal Product Code. All three of these numbers are 6, making the use of the numbers 666 the key to using this identifying marking system." So, as long ago as twenty four years ago, information on the Bar Code's design and numerology was available to the public. However, on his web

(www.http://members.aol.com/product/666ques
t.html) Laurer refutes the claim that his
Universal Product Code is the "Mark of the
Beast". Although he now refuses to answer any
further questions about 666, on his "Questions"
page he initially answered various questions
about the number 666 being part of the UPC by
saying :- "Rumour has it that the lines (left,
middle, right) that protrude below the UPC are
the numbers 666. I typed a code with all sixes
and this seems to be true. Yes, they do resemble
the code for six. There is nothing sinister about
this nor does it have anything to do with the
Bible's "Mark of the Beast." It's simply a
coincidence, like my first, middle and last name
all have six letters." Now of course the big
question is this. Is George Laurer telling the
truth? Has this really nothing to do with 666
and the Mark of the Beast, purely a coincidence
as he mentions? Of course there are many
people who have claimed that the Universal
Product Code does not fit in exactly with the 666
prophecies in the book of Revelation. Although
the number of the beast is 666, they say that in
the two verses preceding verse eighteen of the
thirteenth chapter of Revelation the prophecies
refer to the place where mankind receives the

mark of the beast and to the mark's means of controlling mankind's buying and selling ability in the consumer market.

> *And he causeth all, both small and great, rich and poor, free and bond, to receive a mark in their right hand, or in their foreheads.*

Revelation 13. verse 16.

This verse states quite categorically that the mark will be placed in either our right hand or in our head. Until the end of the twentieth century no-one really knew what this verse meant and what could be given to a human in the right hand or forehead without it being extremely noticeable. The next verse of Revelation tells us a little more about the "mark of the beast" and what control it will have on mankind's capacity to buy and sell in the modern world.

> *And that no man might buy or sell, save he that had the mark, or the name of the beast, or the number of his name.*

Revelation 13. verse 17

Now even though at present we seem to have no need to have a mark placed in our right hand or forehead, the very fact that this verse tells us that no man can buy or sell unless he has the mark of the beast seems to point directly to the Universal Product Code and the Bar Code information contained in each Bar Code, without which no products could be bought, displayed or sold to the public.

More interesting than this is an invention in 2002 which brought the "mark of the beast" and the number 666 directly to the attention of the United States of America and, interestingly, to America's many Christians. As I have mentioned before, the United States is a religiously aware country of some two hundred million Christian believers and the Biblical connotations contained in a new invention announced to the American public in 2002 soon built up a major controversy amongst many of America's Christians, who claimed that this invention was precisely what the Bible was prophesying about the "mark of the beast" and 666 in the book of Revelation.

Applied Digital Solutions (www.adsx.com) is a technology company based in Palm Beach,

Florida, which in 2002 announced its invention of the "VeriChip", a syringe-injectable personal identification and tracking device, about the size of the balltip of a ballpoint pen, that is injected into the right hand palm of its recipient. The "VeriChip" is designed to carry a unique identification number of its user together with other personal details such as bank account details, credit card numbers and medical information. However, the thing that has got so many American Christians annoyed and fearful of this invention is the technology that Applied Digital Solutions uses in the "VeriChip" to store and transmit information of its owner.

Applied Digital Solutions revealed that the technology used in the "VeriChip" was based on the identification technology used in the Universal Product Code or the Bar Code. In other words, VeriChip is using the technology invented by George Laurer whilst he was working for IBM. The VeriChip then, is designed, manufactured and technologically calculated to represent the same numbers as the Bar Code. 666. It is also placed in the palm of the right hand and, having the ability to carry each individual's identity number, bank details and credit card numbers, can therefore be used to buy and sell goods and services, connect with bank details and credit card numbers, as well as identifying each of us through its identification technology. Christians in the United States also say that the ability of "VeriChip" to contain information about an individual's bank account details and credit card numbers could eventually lead to a situation where only those implanted with the "VeriChip" could actually buy or sell goods or use financial, medical or computerised systems and services. They also say that the "VeriChip" enables humans to be financially controlled and their habits monitored by various government agencies and that the Greek rendering for the word "Mark" in the King James Bible translates to the word "etching" in English. The production of the "VeriChip" involves this process in the manufacture of the products circuitry.

If the Bar Code does not represent 666 and the "mark of the beast", which seems unlikely, then surely the "VeriChip" is, without a doubt, the only other product on the planet that can be compared favourably with both the "mark of the beast" and the number 666. Which ever you believe to be the product that characterises the number 666 and the "mark of the beast", there is one piece of evidence that cannot be denied. The mark of the beast and the number 666 are with us now. Everywhere. Fancy a "VeriChip" anyone?

Chapter Ten. The Last Seven Plagues.

*And I saw another sign in heaven, great
and marvellous, seven angels having the
seven last plagues; for in them is filled up
the wrath of God.*

Revelation 15. verse 1

Although this prophecy is from the fifteenth
chapter of the book of Revelation, it is in
Revelation chapter sixteen that the last seven
plagues are described in their fullness. One of
the problems with trying to interpret these
prophecies in the Bible is that they never
seemed to make sense until the end of the
twentieth century. Take the last chapter about
666 as an example. Is this another circumstance
of these prophetic writings only being
understood as the events actually begin to
happen? A warning to us of bad times to come,
perhaps? Well, make your own mind up.

*And I heard a great voice out of the
temple saying to the seven angels, Go
your ways, and pour out the vials of the
wrath of God upon the earth.
And the first went, and poured out his*

vial upon the earth; and there fell a noisome and grievous sore upon the men which had the mark of the beast, and upon them which worship his image.

Revelation 16. verses 1 & 2

So, here we have the first two verses of Revelation sixteen and the description of the first of the last plagues. This "noisome and grievous sore" that is mentioned is a phrase used often to describe plagues, the black death, pandemics and appalling diseases that have occurred in the last thousand years on the planet. It also seems to convey the message that those men who have "the mark of the beast" and worship the image of the beast will be those affected.

What has to be done here is not so much to identify those that are affected by the plague, but to try and work out which problem or disease the first plague is addressing. Earlier in the book I talked about two problems that this first plague could refer too, one already her in the world, the other waiting for its opportunity. HIV-AIDS has so far infected fifty eight million people and more than twenty two million people have been killed world wide. There still seems to be no cure for this terrible disease. The other problem is the eventual emergence of a flu pandemic. If this is a mutation of the H5N1 virus, the British government has said this could kill fifty million people world wide.

So here we have two contenders for the first of the last seven plagues, but that isn't the end of the danger. As I mentioned in chapter six, chemical and biological weapons also pose a huge threat to mankind. Any escape or accidental discharge from one of the world's chemical or biological laboratories could bring a bacteria into the world that would be unstoppable. If these weapons were used by any countries or terrorists to inflict damage anywhere in the world, the end result would be the same. A killer without any effective treatment will have been unleashed. Whether by this time people around the world have been implanted with a "VeriChip" or a Bar Code is impossible to say, but it would seem that mankind faces a frightening future. Whichever of these scenarios comes to fruition, it is clear that the first plague will have huge and terrible consequences on the population of the world.

And the second angel poured out his vial upon the sea; and it became as the blood of a dead man: and every living soul died in the sea.

Revelation 16. verse 3

The second plague is most likely already with us. Reports from NASA's shuttle flights have shown that massive pollution of the world's oceans is clearly visible from space, and the dumping of chemical weapons, nuclear waste and hazardous substances into our oceans over the last sixty years is now giving great concern to environmentalists world wide. It would seem that over the next few years we are going to reap the whirlwind of a failed world environmental policy sown at the end of the second world war. A policy that has seen the industrial world dump its rubbish into our oceans totally indiscriminately, without any regard for the fish and the animals of the sea, and with no thought to the future of the world. So, thanks a lot industrial world.

Interestingly, the forecast that the sea will become "as the blood of a dead man" gives a fairly clear description of the colour the oceans will resemble. Blood that has spilled from someone who is dying becomes a dirty dark brown colour as it dries. I know this is not a particularly pleasant description to provide to the reader, but it is true. As a journalist I have unfortunately seen these terrible circumstances many times and they are not memories and experiences you forget.

Exactly what causes this change of colour to the world's oceans is unclear, but when you look at the scientific information available, particularly from climate change specialists and environmental experts, it seems that the incredible pollution of our oceans is continuing at an unsurpassed speed and there are no government agencies or world bodies prepared to monitor and scrutinise the damage this pollution is causing. Exactly what the second plague is telling us is that as the oceans become more polluted and fouled by the industrial world's dumping of toxins into the oceans, the fish and the animals, as well as mankind, become unable to survive the deadly poisons in the seas, our food from the oceans becomes unfit for human consumption and our ability to use the ocean's waters as a safe and reliable form of irrigation, cleanliness and bathing pleasure becomes impossible. Any idea of desalinating plants converting sea-water to fresh water, a position some governments have advocated as another way of providing the population with clean water, will disintegrate.

What verse three of Revelation Chapter sixteen is telling us is that the waters from the world's oceans are becoming filthy and polluted to such an extent that everything living in the oceans will die and those trying to use the water as irrigation, farming, cleanliness or drinking will suffer catastrophic loss of life. Good news eh?

> *And the third angel poured out his vial upon the rivers and the fountains of water, and they became blood.*

Revelation 16. verse 4

Mankind has, since the industrial revolution of the 19th century, had as much regard for the rivers of the world as for the oceans. Indeed, the pollution of some of the world's major rivers is now such that all life in them is dead, and humans too would die should they drink the water. Not only are we poisoning the rivers, but as with our pollution of the oceans, we are depriving ourselves of one of our most valuable food sources and destroying our ability to use river water for irrigation, drinking and cleanliness. How long will it be, as we continue our pollution, until the fish of the seas and rivers are either dead, or as in many parts of the world already unfit to eat. Not long I fear. As I mentioned in Chapter Two, the World Health Organization (WHO) has estimated that over one billion people do not have access to clean water at the present time, and in the developing nations up to ninety five percent of sewage and seventy percent of industrial waste were simply being dumped untreated into water courses, rivers and lakes, as well as the oceans. This is idiocy on a massive scale that will eventually kill millions.

> *And the fourth angel poured out his vial upon the sun; and power was given unto him to scorch men with fire.*
> *And men were scorched with great heat, and blasphemed the name of God, which hath power over these plagues; and they repented not to give him glory.*

Revelation 16. verses 8 & 9

Once again, a prophecy that seemed to make no sense until climate change and global warming started to change the earth's atmosphere. The ozone layer became depleted and the constant releases of massive amounts of poisonous gasses into the atmosphere has meant that the ultra violet radiation reaching earth has become a major hazard, causing crop failure, blindness and such a huge epidemic of skin cancers and melanomas that it is becoming considered dangerous to expose our skins to the sun for too long a time as our skin will be scorched. The fourth plague is with us right now. As horrific as this seems, there is worse to come.

> *And the fifth angel poured out his vial upon the seat of the beast; and his kingdom was full of darkness; and they gnawed their tongues for pain.*

> Revelation 16. verse 10

This is an interesting prophecy in so far as it seems to indicate the area where this plague originates and that this area will be covered in darkness. Whilst this seems not only highly improbable, but almost impossible, there is in our history a frightening precedent that should be a warning to us all.

In 1159 BC, the volcano Hekla, on Iceland erupted in a massive explosion. This shows up in Irish tree rings of the period as being followed by twenty or so years of extremely low sunshine and heavy rainfall, and it is thought that this darkness extended around much of the northern hemisphere. The effects were disastrous. Crops failed, famine ensued and in the following years the population of Europe was decimated.

In the final part of the verse the words "they gnawed their tongues for pain" is representative of the hunger caused by the famine that inevitably would follow another massive volcanic explosion, and should there be a repeat of the Hekla eruption or something similar, perhaps the eruption of the Yellowstone National Park supervolcano, the dust particles thrown into the atmosphere would not only do tremendous damage to the atmosphere and the ozone layer, but would reflect sunlight away from the earth. The consequent result on the wheat producing areas of the American and Canadian mid-west, the Ukraine and Great

Britain, the vegetable growing areas of northern Europe and America and the fruit growing areas of the Mediterranean, California and France would be totally devastating. Mankind would not be able to feed itself and the resultant chaos is almost too terrible to contemplate.

And the sixth angel poured out his vial upon the great river Euphrates; and the water thereof was dried up, that the way of the kings of the east might be prepared.
Revelation 16. verse 12

This, the penultimate plague, starts by immediately identifying not only the area, but the geographical feature that will be affected. The river Euphrates runs through what was once Mesopotamia, the supposed site of the Garden of Eden. It is the largest river in South West Asia, rising in the foothills of north east Turkey fairly near Mount Ararat. It then flows almost three thousand kilometres south through Turkey, Syria and Iraq, finally disgorging itself into the Persian Gulf through the Shatt-al Arab waterway.

What it is that causes the river Euphrates to be dried up is difficult to say. Global warming and climate change appear to be responsible for many of the world's rivers drying up; just take a look at the Murray Darling river basin, the largest of Australia's river systems, which, due to the worst drought in the country's history has forced the Australian government to refuse farmers in the south east of Australia permission to use the water from the Murray Darling basin for agricultural irrigation. This has caused food shortages and agricultural exports to plummet in a country once rich in agriculture, wheat, fruit and wine.

Is this how the river Euphrates dries up?

Interestingly enough, at the end of first Gulf War in Kuwait and Iraq, Saddam Hussein embarked on a campaign against the Marsh Arabs that involved the draining of the marshes in the Euphrates basin in south west Iraq. Perhaps Saddam Hussein was already beginning to fulfil this prophecy, preparing the way for the "kings of the east".

Who these kings of the east are can only be surmised, but directly to the east of the Euphrates is Iraq, and further east the old Persian empire, now modern day Iran. Further east still, brings us to Afghanistan and Pakistan and onwards over the Himalayas to China. At present there are wars in Iraq and Afghanistan with Iran continuing to pursue their uranium development to the annoyance of the United States, Great Britain, Israel and other powerful countries in the Gulf. Will Iran's continuing expansion of their nuclear production eventually bring about an attack by coalition forces on Iran's nuclear facilities? What will be the result of this? Will this be the justification needed for the way of the kings of the east to be prepared? Who the hell are these kings of the east? What if nuclear weapons are used in the attack on Iran?

To get a clearer picture of this, one has to understand the following facts. Iran is at present being helped by China in its nuclear research and its oil production. It is now known that Iran probably has more oil beneath its homeland than most of the other Middle East countries.

As I have mentioned many times before in this book, energy resources, and in particular oil, has become the most wanted commodity on the planet. Any attack on Iran by western countries would be seen by China as a threat to their oil production and energy resources, something they have worked hard with Iran to facilitate and refine. What would be the effects of a major assault on Iran by a western coalition? Iran may not have nuclear weapons yet, but their army, air-force and navy are strong and very well equipped with radar, missiles and weapons provided by China. China, of course, is a country with a huge nuclear armament and a massive army, navy and air-force.

Pakistan, just to the east of Iran, is also a very powerful nuclear country and may have helped the Iranians with facilities and production work on their nuclear plants and nuclear centrifuges, devices that separate weapons grade uranium from the less exploitable uranium. Pakistan is also the country that passed its nuclear information on to other interested parties. What would their reaction be to an attack on its friend Iran? Will one or both of these countries become confrontational against Iran's attackers? Are they, China and Pakistan, the kings of the east, or does India get involved?

So, the first six plagues move from a terrible disease or pandemic in the world to the possibility of a global nuclear confrontation. Nice. Further information of the six plagues is recounted in the next two verses of Revelation chapter sixteen and they are not pleasant at all.

*And I saw three unclean spirits like frogs
come out of the mouth of the dragon, and
out of the mouth of the beast, and out of
the mouth of the false prophet.*

Revelation 16. verse 13

Some of the prophecies in the sixteenth chapter of Revelation are quite easy to interpret. Others, such as the above, are a little more difficult. Just exactly what are the dragon and the beast? Where and who they might be are an extraordinary problem to understand. Are we talking about these "spirits like frogs" as being men, presidents, dictators or world leaders? Who or what is the false prophet? Just what is going on here? Whoever these "spirits" may be, the next verse of Revelation tells us exactly what their intentions are. To draw the world into its final conflict.

*For they are the spirits of devils, working
miracles, which go forth unto the kings of
the earth and the whole world, to gather
them to the battle of that great day of God
Almighty.*

Revelation 16. verse 14

This is probably one of the most frightening verses of Revelation. Whoever these "spirits" are, their intention is to take the world into a battle that will be utterly inconceivable. The final conflict.

> *And he gathered them together into a place called in the Hebrew tongue Armageddon.*
>
> Revelation 16. verse 16

There has been much speculation about the location of Armageddon. Many have said it refers to a valley in northern Israel close to its border with Syria, whilst others have believed it to be the Bekka valley, further to the east. It is impossible to say with any certainty the precise location of Armageddon, indeed it may not be a place in the standard sense, as some people have translated Armageddon simply as "the final conflict". Whatever the truth, the events that follow are apocalyptic.

> *And the seventh angel poured out his vial into the air; and there came a great voice out of the temple of heaven, from the*

throne, saying it is done.

And there were voices, and thunders, and lightnings; and there was a great earthquake, such as was not since men were upon the earth, so mighty an earthquake, and so great.

Revelation 16. verses 17 & 18

So now we have reached the last of the plagues, the final seventh plague. The thunder and lightning could be synonymous with aerial and nuclear warfare, or massive storms around the world caused by global warming and climate change, but the description of the earthquake is quite conclusive. The phrase "such as was not since men were upon earth" means quite simply the most powerful earthquake ever. So powerful is this earthquake that it wreaks devastation on a world wide scale never seen before. Could this be the Californian earthquake?

Many scientists now agree that the Californian earthquake, when it comes, will not only wreak havoc along America's west coast but would have a disastrous effect in New Zealand and would hit Japan with such force that the islands of Honshu and Hokkaido would be totally devastated. The earthquake that is described in Revelation could be of such a magnitude that not only the Pacific plate is affected, but there could be a major "knock on" effect, causing other great fault lines, the African rift valley running from South Africa through to the Mediterranean, the Anatolia plate running through Turkey and into the Aegean Sea and the Mid Atlantic Ridge to split asunder. An earthquake on this scale would be enormous and cataclysmic. The next two verses of Revelation bring into focus the terrible effects this earthquake will have around the world and to the earth's major cities.

> *And the great city was divided into three parts, and the cities of the nations fell; and great Babylon came in remembrance before God, to give unto her the cup of the wine of the fierceness of his wrath.*

Revelation 16. verse 19

Which "great city" it is that is divided into three parts is hard to interpret. Is this New York, Los Angeles, Rome, London, Jerusalem, Constantinople, Tokyo? Whichever city

Revelation is talking about, it seems obvious that all the cities of the world will be devastated. Not just the cities though. Live on an island anyone? I do.

> And every island fled away, and the mountains were not found.

Revelation 16. verse 20

Recently geologists have discovered that the world's tectonic plates, that part of the earth's crust that the continents are built on, are not only moving against each other, but are extremely fluid. Geologists now believe that because of this the world's greatest mountain range, the Himalayas, could one day collapse under its own weight. As if that isn't bad enough, the islands mentioned could be any islands anywhere in the world. In fact it appears that verse twenty of Revelation chapter sixteen is talking about every island vanishing. How many people would that effect? The answer is millions. Still want to live on your island?

So, there you have the last seven plagues prophesied by the Book of Revelation. From disease or pandemic, through pollution of the oceans and rivers to incredible heat from the sun and eventually the final conflict of Armageddon. The only conclusion we can make from these prophecies is that the last seven plagues are a combination of man-made and natural disasters, concerning our pollution of the planet, man's desire to decimate his neighbour, and natural events we have no control over. There is also one other conclusion we can reach concerning the last seven plagues. They are with us now.

Chapter Eleven. Judgement Day

But of that day and hour knoweth no man, no, not the angels of heaven, but my father only.

Matthew 24. verse 36

Now I realise that this chapter is, to some people, going to be extremely contentious. As I have mentioned before, the prophecies in the Bible are difficult to interpret, but you don't have to be religious or a believer in God to find what the Bible has to say about the future an interesting and thought provoking state of affairs.

The above verse, spoken by Jesus, is from the Book of Mathew in the Bible and is used by many people to indicate the impossibility of anyone knowing exactly when Judgement Day will occur. However, it only talks about the day and the hour of Judgement Day, not the year. So, whilst it is impossible to interpret the day or hour of this event, the Book of Daniel in the Old Testament and the Book of Revelation in the New Testament are able, through simple addition, to give us a series of dates that finishes with Judgement Day.

For hundreds of years mankind has been predicting the end of the world, the second coming, the resurrection, the final judgement. Hardly a decade goes by without at least one group of people announcing the date of the Apocalypse. How these dates are arrived at varies, but very few if any are based upon the predictions and indeed dates that are given in the Bible. Not only does the Bible give the date for this "great change", but as if to confirm that this date is correct, the Bible also gives the three most important dates in the history of the Jewish people since the destruction of the second temple in 70 AD. But, for over a thousand years the interpretations of these dates have been incorrect, due to two inherent errors made almost two thousand years ago. Indeed, the key to this mystery goes back almost to the seventh century BC, a key which when used to interpret both the Old and New Testament prophecies, reveals the Bible's greatest secrets.

First however, two mistakes in biblical interpretation have to be identified and rectified before we progress any further. The first mistake concerns an event that ranks as one of the most important in the history not only of the Jewish faith, but of Christians and Muslims alike. It is the "abomination of the desolation."

> *And arms shall stand on his part, and they shall pollute the sanctuary of strength, and shall take away the daily sacrifice, and they shall place the abomination that maketh desolate.*

> Daniel 11. verse 31

In "A Dictionary of the Bible" published at the beginning of the twentieth century, James Hastings M.A. D.D. analyses the different interpretations of the above verse and concludes that it refers to:- "The setting up by Antiochus Epiphanes of a small idol/altar on the altar of the Holy Temple in Jerusalem in 167 BC". This refers to the invasion and capture of Jerusalem by the Seleucids under Antiochus in 167 BC, an event that spawned the Maccabean revolution, which after the death of Judas Maccabeus in battle in 160 BC was successful in expelling the Seleucids from Jerusalem. To this very day theologians and scholars alike have accepted this interpretation of the "abomination that maketh desolate". However, they are wrong. Daniel the sixth century prophet, is not the only place in the Bible where this event is mentioned. It is also mentioned emphatically in two of the New Testament gospels and is spoken about by Jesus Christ himself.

> *When ye therefore shall see the abomination of desolation, spoken of by Daniel the prophet, stand in the holy place, (whoso readeth let him understand).*
>
> St Matthew 24. verse 15

> *But when ye shall see the abomination of desolation, spoken of by Daniel the prophet, standing where it ought not, (let him that readeth understand).*
>
> St Mark 13. verse 14

These two verses allude to the conversation between Jesus, Peter, James, John and Andrew on the Mount of Olives shortly after Jesus has admonished the Pharisees and Sadducees in the temple, just two days before the feast of the Passover. Before he mentions the "abomination"

Jesus says the following to his disciples, who are admiring the grandeur of the temple.

> *Seest thou these great buildings? there shall not be left one stone upon another, that shall not be thrown down*

St Mark 13. verse 2

Here, at the beginning of Mark chapter thirteen, Jesus is prophesying an event that is to take place within the lifetime of the disciples. In 70 AD the Roman Legions under Titus Andronichus, son of Emperor Titus Flavius Vespasian, on the orders of Nero before his death, captured and destroyed Jerusalem, in doing so completely razing the Temple of the Jews to the ground., leaving only the "Western" or wailing wall standing. As Jesus places the event of the "abomination" after the destruction of the temple in 70 AD and also uses the words "when ye shall see" (Mark 13 verse 14), therefore placing this event in the future tense, it is impossible that the "abomination that maketh desolate" took place in 167 BC. However, there are many clues that point to the correct date, the first of these being in the previous gospel verses.

The inclusion of the words in brackets in Matthew 24 verse 15 and Mark 13 verse 14 are a clear indication that the reader should pay very great attention to the verse and in particular the sentence immediately before the brackets. Therefore the words "stand in the holy place" (who so readeth let him understand) and "standing where it ought not" (let him that readeth understand) must be considered as the most important part of these verses. What both verses are saying is that the "abomination" will be placed somewhere that is not only a "holy place", but also a place or position that under normal circumstances would be forbidden to it. Within the Jewish faith the words "holy place" means only one thing, the Holy Temple in Jerusalem.

After its destruction in 70 AD the Holy Temple was never rebuilt. The ground on which the Temple had stood, the Temple Mount, remained beyond the Western or Wailing wall as the holiest place of Judaism. However, it was not to remain so. In 638 AD, six years after the death of the Mohammed (PBUH), Abu Bakr the first caliph of Islam, invaded and captured Jerusalem. On the death of Abu Bakr, Omar ibn al Khattab, advisor to Mohammed, became the

172

second caliph. It was on Omar's orders that a magnificent golden domed mosque was built on the site of the old Temple. The construction of Islam's third holiest shrine, standing directly over the holiest place of the Jews on the Temple Mount began in 688 AD. The Mosque of Omar ibn al Khattab is the Dome of the Rock. This is the true identification of the "abomination that maketh desolate". From 688 AD the Jewish people were not only without a homeland but they had been denied access to their holiest place of worship. As a people and a nation they were indeed desolate.

Having identified and rectified the first interpretative error in the Bible, it is time to search for the second. This time we are not looking for a starting date by which to measure our timings, but at the timings themselves and the fact that the timings give us the key to these extraordinary prophecies. A key that when turned gives us the most incredible set of dates.

> *For I have laid upon thee the years of*
> *their iniquity, according to the number of*
> *the days, I have appointed thee each day*
> *for a year.*
>
> Ezekiel 4. verses 5 & 6

In these two consecutive verses from the Old Testament book of Ezekiel, we are given the key to unlocking the Bible's greatest secrets. Count the days as years. Simple. The instruction that each day should count as a year, though given to Ezekiel specifically, would have been understood by the other biblical prophets as applying to them also. Daniel, a keen scholar and student of Ezekiel would have seen this instruction as applying to himself. Writing some thirty years after Ezekiel the words "I have appointed thee each day for a year" were to Daniel, the true word of God and as such became constant.

> *How long shall be the vision concerning*
> *the daily sacrifice, and the transgression*
> *of desolation, to give both the sanctuary*
> *and the host to be trampled under foot.*
> Daniel 8. verse 13

Daniel chapter 8 is one of the most illustrative chapters in the Bible, giving the reader a full and comprehensive description of events that are well documented in history and one incredible date that is yet to come. Daniel is given these dates in a vision that is explained to him in a very explicit and unambiguous way. It is this description that enables the reader to place the events in the chapter quite conclusively, especially considering that "each day be counted as a year". Now it becomes very interesting.

> *And he said unto me, Unto two thousand and three hundred days, then shall the sanctuary be cleansed.*
>
> Daniel 8. verse 14

Here Daniel is told that the two thousand three hundred days marks the time that Jerusalem, the sanctuary and the host, the Temple Mount, will be under the domination of the Gentiles or non Jews. They will "trample it under foot." Daniel is also told that the starting date for this period is the beginning of his vision. The vision that Daniel is given in chapter 8 is extremely specific and starts with one of the greatest events of pre Christian history.

> *Then I lifted up mine eyes, and saw, and, behold, there stood before the river a ram which had two horns: and the two horns*

were high; but one was higher than the
other, and the higher came up last.

Daniel 8. verse 3

Further on in the same chapter Daniel is given an explanation of this verse.

The ram which thou sawest having two
horns are the Kings of Media and Persia.

Daniel 8. verse 20

The first Persian King to also be King of the Medes was Darius I (548 - 486 BC). The last of the Persian Kings was Darius III, defeated in one of the most significant battles in history. Further information as to a starting date for the vision is then given.

And as I was considering, behold, an he
goat came from the west on the face of the
whole earth, and touched not the ground:
and the he goat had a notable horn
between his eyes.

Daniel 8. verse 5

Once again a further explanation is given.

> *And the rough goat is the King of Grecia;*
> *and that great horn that is between his*
> *eyes is the first King.*

Daniel 8. verse 21

The clues to these verses are in the words "and touched not the ground" and the single word "great". We have identified the ram as being the Persian Empire. The goat signifies the Greek Empire led by Alexander the Great, the first commander to lead his army on horseback, his feet not touching the ground.

> *And he came to that ram that had two*
> *horns, which I had seen standing before*
> *the river, and ran into him in his fury*
> *and his power.*

Daniel 8. verse 6

What Daniel is seeing in his vision is a battle, between the Persian Empire under Darius III and the Greeks led by Alexander the Great, which takes place across a river. There is only one battle this could possibly be. It is the Battle of Issus, which took place across the River Penaris on what is now the Mediterranean coast of south eastern Turkey, just north of Syria. In the battle Alexander defeated the Persian army and their leader King Darius III. The date of this battle was 333 BC. Taking this as the starting date of the vision and considering Ezekiel's instruction to count every day as a year, adding two thousand three hundred years to 333 BC gives us the year 1967.

On June 5th 1967 Israel launched a lightning attack on its neighbours at the beginning of the Six Day War. In the war the Israeli army surrounded and sealed off Jordanian owned East Jerusalem and claimed it as Israeli territory. For the first time since the capture of Jerusalem and destruction of the temple by Titus in 70 AD, the people of Israel not only had their own state but once again controlled their holiest of holies, the Temple Mount.

There are two further dates that are given in the Book of Daniel and these must be looked at applying the same rules.

And from the time that the daily sacrifice
shall be taken away, and the abomination
that maketh desolate set up, there shall be
a thousand two hundred and ninety days.

Daniel 12. verse 11

In the first part of this verse we are given the starting date of 688 AD, the date that the "abomination that maketh desolate" was set up. If you add to this date one thousand two hundred and ninety "years" instead of days we arrive at the year 1978.

On September 17th 1978 after months of negotiation, one of the most extraordinary events of the twentieth century took place. President Jimmy Carter of the USA, Egyptian President Anwar Sadat and Israel's Prime Minister Menachem Begin signed the Camp David Agreement. For the first time since the Exodus from Egypt in 1290 BC, Israel was at peace with its Arab neighbour.

So, The Book of Daniel has now given us two of the most important events is the history of the Jewish faith and has one more date in the future to be considered. However, it is the Book of

Revelation that gives us the third historically important date, probably the most remarkable event in the history of Israel.

> *But the court which is without the temple leave out, and measure it not; for it is given unto the Gentiles: and the holy city shall they tread under foot forty and two months. And I will give power unto my two witnesses, and they shall Prophesy a thousand two hundred and three score days, clothed in sackcloth.*

Revelation 11. verses 2 & 3

The first thing that seems clear in these two verses is that they appear to be talking about the same time scale. If an average month is thirty days, then the forty two months of verse 2 coincides exactly with the one thousand two hundred and sixty days of verse 3.

The other two facts we know from these verses is that in verse 1 the area outside the temple will be in the hands of the Gentiles, i.e. non Jewish people, as will the holy city of Jerusalem for that amount of time; and in verse 2 the two witnesses of God, in other words the Houses of

Judah and Israel, the Jewish people, will prophesy or preach for a similar period "clothed in sackcloth." Sackcloth is symbolic of regret and repentance and the ancient custom of wearing sackcloth signified penitence and mourning.

What these two verses are saying is that the Jewish people will lose their place of worship, including the city of Jerusalem; it will be given over to non Jewish people for a specific period of time. This coincides with a similar period of time during which the Jewish people are in mourning not only for the loss of their place of worship, but of their homeland.

And the woman fled into the wilderness, where she hath a place prepared of God, that they should feed her a thousand two hundred and threescore days.

Revelation 12. verses 5 & 6

Once again one thousand two hundred and sixty days are mentioned here and refer to a time spent in the wilderness. The woman referred to, who is to bring forth a manchild who will rule all nations with a rod of iron is not only symbolic of Mary the mother of Jesus, but of the entire Jewish nation. Once again these two verses prophesy that the Jewish nation will wander in the wilderness for one thousand two hundred and sixty days.

Saint John the Divine, writing the Book of Revelation, would also have been aware of the instruction that each day should be counted as a year and so the "one thousand two hundred and threescore days" of Revelation chapters 11 and 12, become years. Taking the dates of the "abomination that maketh desolate" as the beginning of the period of Jewish desolation, we have 688 AD. If one adds one thousand two hundred and sixty "years" to this date we arrive at the year 1948.

On May 14th 1948 the British mandate for Palestine ended and the State of Israel was created. For the first time since 688 AD the Jewish faith once again had a homeland and the State of Israel became a reality.

So now the Bible has correctly given us the three most important dates in the history of the Jewish people in the twentieth century; 1948, 1967 and 1978. There is however, one more date given within the pages of Daniel, a date yet to come, a date of enormous significance.

Blessed is he that waiteth and cometh to the thousand three hundred and five and thirty days.

Daniel 12. verse 12

The interesting word in this verse is "Blessed". Although used often in the Bible, it is a word which is always descriptive of a state of grace that each individual can attain, a oneness with God, a state of redemption.

Once again we must obey the instruction laid down by Ezekiel and count those days as years and again the starting point for our calculations is the "abomination that maketh desolate", in other words 688 AD. Adding one thousand three hundred and thirty five years to this date gives us the year 2023. Having been given this final dating Daniel is then told, in the last verse of his book:-

But go thy way till the end be: for thou shalt rest, and stand in thy lot at the end of the days.

Daniel 12. verse 13

The finality of this last verse is unequivocal and relates directly to the penultimate verse stating as it does that those who reach the "one thousand three hundred and five and thirty days" are not only "blessed" but will have reached the "end of the days". That the Bible is correct in its first three datings of 1948, 1967 and 1978 cannot be in doubt. Who then can doubt that the final dating given is not also correct? It is a date that immediately follows Armageddon: Judgement Day, the day of Resurrection. A cataclysmic event that will change the world. The date is 2023. Of course, what the Book of Daniel doesn't tell us is what kind of events are happening on the planet before Judgement Day. However, there are other parts of the Bible, particularly in the New Testament, that do seem to give a description of a scenario that may occur as we approach Judgement Day. A scenario that has some very frightening implications for both the planet and the population. First though, a final word from the Book of Daniel about the coming apocalypse.

And at that time shall Michael stand up, the great prince which standeth for the children of thy people; and there shall be a time of trouble, such as never was since there was a nation even to that same time: and at that time thy people shall be delivered, every one that shall be found written in the book.

Daniel 12. verse 1

So here Daniel is talking about the appearance of the Archangel Michael at a time of great trouble, greater than any trouble the planet has ever witnessed, that finishes with a judgement on the people. This description in Daniel is almost identical to that given by Jesus to his disciples when they ask him what is to happen in the "latter days."

For then shall be great tribulation, such as was not since the beginning of the world to this time, no, nor ever shall be.
And except those days should be shortened, there should no flesh be saved: but for the elects sake those days shall be shortened.

Matthew 24. verses 21 and 22

Once again, these verses talk about a time of

great trouble, worse than anything ever seen before in the world. The verse also states that this trouble has to be ended somehow or no human beings would be left alive. What is interesting is how Jesus then describes the troubles to his disciples further on in Mathew 24 and in the Books of Mark and Luke. The meaning of what he says has, however, been obscured by an imperfect translation in the Latin or Vulgate Bible as well as the Saint James version of The Bible. Check this out.

Immediately after the tribulation of those days shall the sun be darkened, and the moon shall not give her light, and the stars shall fall from heaven, and the power of the heavens shall be shaken.

St. Matthew 24. Verse 29

And the stars of heaven shall fall, and the powers that are in heaven shall be shaken.

St. Mark 13. Verse 25

Men's hearts failing them for fear, and for looking after those things which are coming on the earth: for the powers of heaven shall be shaken.

St. Luke 21. Verse26

To gain the full significance of these three verses, it is necessary to look at how the original texts in the New Testament of the gospels were written. The language of the apostles writing these texts was Greek, indeed the Old Testament containing 39 books was written mainly in Hebrew, with some Aramaic, but first collected together and translated into Greek in the third century BC. This was known as the Septuagint Old Testament. The writings of the New Testament were collected together in the second century AD in the original Greek and not until the fourth century AD did the Latin or Vulgate Bible appear.

The three verses mentioned previously all refer to the "power of the heavens being shaken". Taken literally this seems to be an event that is hard to understand. Mention of the sun and moon being darkened could be explained in many ways. Pollution spreading into the atmosphere or volcanic eruptions could cause just such an effect. As could a nuclear war. That the "heavens could be shaken" however seems extraordinary until you look at the original Greek text. The word for the "Heavens" in Greek is "Ouranus". Herschel took this word in March 1781 to name his newly discovered planet Uranus. Eight years later in 1789, a mining engineer discovered a white metal shining out from the black rock-face he was cutting. He thought the white metal particles looked like the stars in the heavens. The mining engineer was named Klaproth, the metal element he had discovered he called Uranium.

Is the real message contained in the writings of Matthew, Mark and Luke that the "power of uranium shall be shaken". In this context it could mean only one thing - nuclear war. Finally, Jesus follows this with a parable that gives a quite specific time period for the events he has spoken of earlier.

Now learn a parable of the fig tree; when his branch is yet tender and putteth forth leaves, ye know that summer is nigh:
So likewise ye, when ye shall see all these things, know that it is near, even at the doors.

Saint Matthew 24. verses 32 & 33

In Biblical terms the fig tree is always symbolic of Israel. As we have seen earlier in the chapter the State of Israel came into being on May 14th 1948. What this verse is alluding to is a young State of Israel, its population spreading across the country and becoming fruitful. Since 1967 and the six day war, this is indeed what Israel has done, moving as it has into the West Bank, Gaza Strip, Sinai and the Golan Heights and creating new settlements, turning arid desert into fertile land. The warning given in verse thirty three is very specific, saying as it does that when the world sees Israel growing as a nation, then the tribulations he has forecast will come about. The next two verses are even more specific and give a time scale for these events that coincides with the other prophetic timings perfectly.

> *Verily I say unto you, this generation shall not pass, till all these things be fulfilled. Heaven and earth shall pass away, but my words shall not pass away.*

> Saint Matthew 24. verses 34 & 35

What Jesus is saying here is simply that the generation that sees the State of Israel come into being and start to expand, will be the generation that witnesses the terrible tribulations he has spoken about. Those people therefore, born between 1946 and 1964, the "baby boomers", are the generation that will witness the most tremendous upheaval mankind has experienced since the great flood. A time of natural disasters, revolution, famine and nuclear war on an unprecedented scale and at the finish - Judgement Day.

The message Jesus gives us in chapter 24 of Saint Matthew's gospel tells us who will witness this event. It is an event that will be seen by that generation of mankind that has seen the formation of the State of Israel. We are that generation.

It is also interesting to note that both the ancient Egyptian and the Mayan civilisations predict that the earth and its population will suffer major problems and unrest which seemingly start at the beginning of twenty first century. The Great Pyramid of Cheops in Egypt has been studied by archaeologists, scientists, academics and Egyptologists for thousands of years with some remarkable conclusions. The passages and chambers within the Pyramid appear to represent a timescale that some commentators believe gives a clear set of datings for events in the future. Amongst those who have written at length on this matter are Morton Edgar and Max Toth, both of whom use measurements devised by John Davidson to date the chambers in the Pyramid.

The conclusion they have reached involves the Great Subterranean Chamber in Cheops, and what most archaeologists call the "Pit of Ordeal" or the "Pit of Destruction." Though datings are a little imprecise, about + or - 3 years, there can be no doubting that by using Davidson's method of measurement, namely one primitive inch equals one year, the dates 1914 and 1939, representing World Wars 1 and 2 are clearly indicated, and using the same method, the Great Subterranean Passage enters the "Pit of Destruction" in 2003 and seems to last some twenty years before coming to a dead end.

The Mexican Mayan civilisation, now proved to be closely connected with the Egyptians, also appears to give a dating for what seems to be a major catastrophe on earth. Archaeologists are unsure if this dating pertains to the end of the world, or to an event that is the precursor of the end time. The date, however, is extremely precise. December 24th. 2012. Many readers will find these datings spurious, but they are well publicised and have one thing in common which is undeniable. They point to a major world wide calamity at the beginning of the twenty first century.

Whether you believe in the Bible or not, the descriptions contained in it, particularly in the Books of Daniel and the Revelation, do seem to have parallels with events that are taking place now, at this very moment. The Book of Daniel also seems to give us the final date, Judgement Day. As I mentioned before, the date is 2023. Not long now eh?

Perhaps, therefore, we should be afraid, very afraid, of what comes next.

Chapter Twelve. Conclusion.

"None are more hopelessly enslaved than those who falsely believe they are free."

Johann Wolfgang von Goethe (1749 - 1832)

Well, that's the Biblical prophecies and other predictions sorted out, so here we are at the last chapter of the book and I can hear you thinking: "Maybe he's going to tell us some good news now."

Well I'm sorry, but you're wrong. Let me liken this last chapter to the patient who visited his doctor to be told :- "I have some bad news and some worse news for you, which would you like first?"

"You better tell me the bad news first." said the patient.

"You have twenty four hours left to live." said the doctor.

"What's the worse news?" said the patient.

"I should have told you yesterday." replied the doctor.

So, that's the way it's going to be in the last chapter. More bad news. I first started writing this book in March 2008 and now here we are twelve months later and much has happened between my starting the book and concluding it. First though, let's just deal with the economic crisis. This problem is something that, until it happened, was denied by just about every banker, politician and moneyman in the western world, particularly in America. Don't forget that so far, the United States has apparently spent $5000000000000 on the war in Iraq. In case you don't understand that financial figure, let me explain that the figure is five trillion dollars. Five years of war, one trillion dollars a year. All of it is taxpayer's money. Anyone winning yet, you think?

Now the United States government has decided to place $700 billion of the American taxpayer's money into the troubled banks. Hang on, aren't banks supposed to be honest and prudent with the taxpayers invested savings? Aren't banks supposed to be the careful and diligent in using their investor's money properly? Indeed, as countries like Iceland, Ukraine, Hungary and South Korea go cap in hand to the International Monetary Fund (IMF), some economic commentators are saying that taxpayers around the world could eventually fork out three thousand trillion dollars to keep the world economically viable. This sounds a bit crazy doesn't it?

Now ask yourself this question. Are the stock markets around the world reacting positively to this financial help? No. The economies of most countries, along with their banks and businesses, are collapsing. What an unqualified mess the banks and the politicians have got the planet into.

Most analysts are now predicting that the world will go into recession in 2009 and it will last for three years. I don't want to go into the details of what a recession will cause around the world, but it would seem that most people are pretty scared of what is to come. The only unfortunate consequence of this is that most people will be more concerned about the economy than they will be about global warming, climate change and sea level rise, let alone nuclear war. One thing, however, is for certain. A recession causes massive problems around the world and doesn't last for just a few months.

The economic problems should not, however, conceal the environmental problems that have come to the attention of scientists over the last seven months. First, let's deal with a serious problem that is happening underneath the Arctic Ocean. Yes, you read me correctly, underneath the Arctic Ocean. As I said at the beginning of the book, the Arctic is going through rapidly changing conditions at the present time, with water temperatures, atmospheric temperatures and massive ice melt starting to change the entire climatic cycles and meteorological conditions of the area. In the last thirty years the average temperature in the Arctic region has risen by four degree Celsius and air temperatures over the Arctic region increased by two degrees Celsius between 1978 and 2006.

One of the main research projects in the Arctic is being carried out by the United States National Snow and Ice Data Centre (NSIDC) in Boulder, Colorado. They found that in the summer of 2007 the Arctic sea ice shrank to the smallest extent ever recorded. In 2008 the scientists predicted that the Arctic would be ice free by the summer of 2013. That's only four years away.

Doctor Julienne Stroeve, a research scientist from the NSIDC, says :- "The real issue in the Arctic now is that most of the pack ice has become really thin and a regular summer, with high temperatures like 2007, would just melt the ice away."

Another scientist, Doctor Ian Willis from the Scott Polar Research Institute in Cambridge, sees a very frightening scenario beginning to develop as the Arctic ice starts to melt and eventually vanish. He comments that :- "Sea ice has a higher reflectivity than ocean water; so as the ice melts, the water absorbs more of the Sun's energy and warms up more, and that in turn warms the atmosphere more, including the atmosphere over the Greenland ice sheet. Greenland is already losing ice to the oceans, contributing to rises in sea levels. The Greenland ice cap holds enough water to lift sea levels globally by about seven metres, that's twenty two feet, if it all melts."

Now work that out. Live in a city by the ocean anyone?

Of course, earlier in the book I mentioned the fact the Kangerdlugssuaq Glacier on the east coast of Greenland is melting extremely quickly. The glacier deposits tens of cubic kilometres of fresh water into the North Atlantic, its daily fresh water loss being equivalent to the total yearly water consumption of New York City. Doctor Gordon Hamilton, of the Climate

Change Institute at the University of Maine, one of the scientists studying the glacier says :- "The predictions for both the rate and the timing for sea level rise in the next few years have been hugely underestimated."

So it appears that sea level rise may well increase far more quickly than even the scientists had predicted and the ramifications of this sea level rise to the entire world will be catastrophic. Forget about a seven metre rise in the oceans, just consider how a one metre rise would affect some of the biggest and most populated coastal cities in the world. I've lived in Newcastle upon Tyne, Liverpool, New York, Los Angeles, Amsterdam, London and a small Greek island in the Cyclades. If the sea level rose by just one metre, all these cities would be underwater and the small island in the Cyclades would probably become uninhabitable. I suggest that if you want to visit the Maldives in the Indian Ocean, for instance, or any other low lying islands in the world, do it now before they vanish. Forever.

In September 2008, amidst all the other warnings from scientists about climate change, global warming, sea level rise and the melting of the Arctic ice, a new threat emerged that has really started to worry the scientists who discovered it. As the Arctic ice melts and more of the sun's energy is absorb by the ocean, the ocean itself is getting warmer and the permafrost of the Arctic seabed is beginning to melt and become perforated. This has started to form small holes in the permafrost that is letting the most dangerous of greenhouse gases escape and rise up into the atmosphere. Million of tons of this greenhouse gas, which is twenty times more powerful than carbon dioxide (CO_2), are being released. The gas is Methane.

Doctor Igor Semiletov, from the Far Eastern branch of the Russian Academy of Sciences (www.ras.ru) this year oversaw the International Siberian Shelf Study aboard the Russian research ship Jacob Smirnitskyi. Since 1994 Doctor Semiletov has led ten expeditions in the Laptev Sea and has covered thousands of square miles of the Arctic Ocean and the East Siberian continental shelf. In September 2008 the scientists aboard the research ship discovered a large area of the Arctic Ocean that was releasing immense amounts of methane into the atmosphere.

The release was so intense that the methane did not have time to dissolve into the sea water and investigation by echo sounders and seismic instruments found that the methane was bubbling through methane chimneys rising from the sea floor. Millions of tons of methane is being released into the atmosphere and scientists now fear that these huge releases will accelerate global warming, with the atmospheric methane causing higher temperatures in the Arctic, causing further melting of the permafrost and therefore releasing even more methane into the atmosphere. It seems that we may have just witnessed a "tipping point" underneath the Arctic Ocean which we cannot stop.

Most scientists now agree that the amount of greenhouse gases in the atmosphere has exceeded four hundred parts per million. At this point the global temperature rises by two degrees Celsius. However, the release of so much methane into the atmosphere will exacerbate global temperatures dramatically, affecting the entire planet. As the ice sheets melt in Antarctica, allowing its glaciers to flow into the sea, and the Greenland glaciers go into irreversible meltdown in the Arctic, the effect on sea level rise will be astronomical, displacing possibly two billion people, making many parts of the planet, including major cities, uninhabitable and bringing water shortages and food rationing to many parts of the world. The population of Earth will begin to starve. This is not good news.

As the methane starts to do its damage to the planet, with mankind unable to do anything to stop the huge releases of the gas into the atmosphere, it's as well to mention another big problem the Earth has, that politicians and world leaders are still not taking any notice of. This is something that has been going on now for many years. Everybody knows it is happening. Everybody knows what damage it is doing to the planet. Everybody knows that if we don't stop this destruction it will bring about mankind's extinction. Strangely though, nobody cares. If you don't care and do nothing, you will die. I hope all you people out there reading this are now a little embarrassed about your uncaring attitude to the world's population and the future of the planet. Hang your heads in shame, you stupid people.

The world's tropical rainforests are being destroyed at an astonishing rate and the details of this destruction and its effects on the Earth are only too well known. However, nobody is doing anything about it. The deforestation of the rainforests was not even included in the Kyoto protocol and is outside the carbon markets that the Intergovernmental Panel on Climate Change (IPCC) pointed out in May 2007 was the best hope for halting catastrophic global warming.

Fifty million acres of rainforest, that's the size of England, Wales and Scotland, is felled annually in Brazil, the Congo, Indonesia and elsewhere. However, it is the release of carbon dioxide that is the most frightening aspect of this deforestation. Every day deforestation releases as much carbon dioxide into the atmosphere as eight million people flying between London and New York. Every year deforestation releases two billion tons of carbon dioxide into the atmosphere. More than fifty per cent of life on Earth is in the tropical rainforests, which actually only cover seven per cent of the planet's surface and tropical rainforests also generate the majority of rainfall world wide and act as a thermostat for the Earth. The tropical rainforests are incredibly important to life on the planet.

In May 2007 Indonesia became the third largest emitter of greenhouse gases in the world. Close behind Indonesia is Brazil. Neither of these two countries has heavy industry on a comparable scale with the European Union, India or Russia and yet their carbon dioxide emissions are far larger than any other country in the world except the United States and China. Indeed, smoke stacks climbing into the sky from both countries is very easily seen from space and Sir Nicholas Stern, the author of The Stern Report, presented to the International Climate Forum in Bali in November 2007, has said that :- "The destruction of the rainforests in the next four years will pump more carbon dioxide into the atmosphere than every flight in the history of aviation to at least 2025."

The facts are quite staggering. The Global Canopy Programme, a group of leading rainforest scientists based in Oxford in the United Kingdom, released in 2007 a report that states that the destruction of the rainforests accounts for twenty five per cent of the global emissions of greenhouse gases every year, whilst transport and industry account for fourteen per cent each and aviation only three per cent of the world's total carbon dioxide releases.

It also has to be understood that the rainforests of the Amazon, the Congo and Indonesia are the lungs of the planet. The trees breathe in carbon dioxide and change this to oxygen, which is then released back into the atmosphere. When a tree is cut down in the rainforest, its ability to convert carbon dioxide to oxygen finishes and all the carbon dioxide contained within the tree is released back into the atmosphere, together with the CO2 released as the forest is burned.

Perhaps the most compelling shock about the world's rainforests, however, is the fact that what remains of the standing rainforests contains one thousand billion tons of carbon dioxide. That is double what is already in the atmosphere. If we the people, along with the politicians and world leaders continue to do nothing about the destruction of the tropical rainforests, within the next few years the results of our stupidity will become quite clear. Mankind will start to become extinct.

As British philosopher Bertrand Russell said in his 1993 book The Triumph of Stupidity :- "The fundamental cause of all our troubles is the fact that in the modern world the stupid are cocksure while the intelligent are full of doubt."

I told you we were stupid, didn't I.

So far in the final chapter I have written about the methane rising from the Arctic sea floor and the huge amounts of carbon dioxide (CO2) being released from the destruction of the tropical rainforests in the Amazon, Congo and South East Asia. Of course, there is nothing anyone can do about the massive releases of methane from the Arctic sea. No politicians, scientists, world leaders or members of the general public are able in any way to stop the methane from escaping through the permafrost. It will continue and get worse - and all the scientists know this to be a fact.

However, there are many things that governments, countries, politicians and world leaders can do about the destruction of the tropical rainforests. The big problem is that no-one, not a single politician, not a single world leader, not one single country is doing anything to halt this massive release of carbon dioxide into the atmosphere. Why?

Hylton Phillipson is a trustee of Rainforest Concern and explains why no action is being taken about the rainforests by saying :- "In a world where we are witnessing a mounting clash between food security, energy security and environmental security, while there is money to be made from food and energy and no income to be derived from the rainforest, it's obvious that the rainforest will be ignored."

This does rather beg the question "Are politicians only interested in money?" To a great extent the answer has to be yes. Just take a look at the financial situation of many politicians around the world these days and the clarity becomes apparent. Though many will at first claim to be "only interested in my constituency;" as their tenure of the party ticket gets longer, they begin to get sidetracked by the huge amounts of money politicians can make if they join the business community. Once a politician has joined a business company, then the idea that he or she is transparent, honest, trustworthy, sincere, straightforward and still has his constituents' interests at heart falls by the wayside. Once a politician joins the business community, their ideals, morals and principles vanish, to be replaced with greed, avarice and self indulgence. They have ceased to be an honest politician and have become a power wielding business person whose votes can bring good attributes to the business corporations they represent. They are political whores.

This whole disgusting mess was aptly described by the former American president Theodore Roosevelt as long ago as 1906 :- "Behind the ostensible government sits enthroned an invisible government owing no allegiance and acknowledging no responsibility to the people. To destroy this invisible government, to befoul the unholy alliance between corrupt business and corrupt politics is the first task of the statesmanship of today."

Although this was written over a hundred years ago, it remains specific and unambiguous to this day. Even great dictators in the latter part of the twentieth century were only too aware of the powerful and influential hold that business corporations have on politicians and the political pragmatism within a country.

> "Fascism should more appropriately be called Corporatism, because it is a merger of State and corporate power."
>
> Benito Mussolini

If you want a reason why no world leaders, presidents, prime ministers and politicians are doing anything to stop the destruction of the rainforests and the huge amount of carbon dioxide (CO_2) being released into the atmosphere, here's your answer. Happy? I hope not.

Over the course of this book I have attempted to provide the reader with the latest information from scientists, organisations and government bodies that are continually monitoring, measuring and analysing the effects that ice melt, sea level rise, global warming, climate change, food and water shortages and energy depredation will have on the planet and the population. It's not a good story. When I first started to interest myself in the environment some thirty years ago, most of the people talking about the problems the world may face if no action was taken where looked down upon, laughed at and humiliated, many of them being described as "merchants of doom." Unfortunately, these "doomsayers" appear to have been proved correct. Not only that, but in the past two years organisations and scientists around the world are beginning to admit that their projections for sea level rise, climatic change and global warming were much too conservative.

Not so long ago they anticipated that the Earth would warm slowly with sea level rise being a problem that we would face in 2050. Wrong! They said the West Antarctic ice sheets, particularly the huge Wilkins ice shelf, would remain stable until the 2020's. It's broken up already! They predicted that the Arctic ice was unlikely to melt completely and Greenland's glaciers would not melt into the Arctic Ocean. Oops!

Fortunately, these scientists are clever people and are ready to admit their mistakes regarding the exponential consequences for the environmental eco-systems of the planet and the speed that ice melt, sea level rise, global warming and climate change will influence the entire world. The science behind these global changes has improved tremendously over the last few years, enabling a far more robust understanding of how eco-systems interact with each other. This has allowed the scientists to make far more accurate calculations as to the rate and speed of climate change, ice melt, sea level rise and global warming. As I have mentioned before, none of this is good news.

What we seem to have here is a situation that will cause massive world wide disruption. As the planet heats up, rainfall increases and sea levels rise we are heading for a catastrophe unimaginable a few years ago. The big problem here is that it is going to happen sooner than we expected. According to the scientists, much sooner. The devastating consequences of these global changes will NOT start to happen slowly over the next forty years. They will start to occur rapidly, in perhaps the next five or six years.

Sharp rises in global temperatures, huge increases in rainfall, enormous amounts of glacial ice melting into the Arctic and Antarctic oceans and the corresponding rise in sea levels around the planet, will change the way that mankind views life on Earth. Suddenly life will become a question of survival. How mankind will react to this change is a question we should all be asking ourselves, because we are the ones who have to change. What will you do?

Despite the dire warnings of world-wide calamity, possibly on a Biblical scale, there are still many people in the world today who either do not believe this information or who, for reasons we can only speculate about, choose to ignore these warnings. It is obvious that many of these people have children and grand-children of their own whom they wish to see prosper in a peaceful and safe world. Yet, by their sceptical attitude to these scientific warnings of catastrophe, they bear a great responsibility should their offspring have to face a future legacy of an unsafe and frightening planet. Why? What is it in the nature of man that makes him ambivalent to the suffering of others, whether now or in the future? And for how long has man been like this? Recently, or since the beginning of time?

Evidence in the history of the human race seems to point to the fact that mankind has, since he first walked out of the jungles of Africa, had a deep, almost in-bred predisposition to destroy. Look at recent evidence. Not content with the murder of over one hundred million people during the twentieth century, through wars, programs, despotism, dictatorship, and oppression; it seems that at the beginning of the twenty first century mankind has now embarked on a premeditated course that could end up seeing the human race wiped from the face of the planet. Is this really why we're here? I think not. Yet when we look around the world at the conflicts, disputes, and atrocities that man continues to perpetrate on his fellow man, together with the unending destruction of our ecology and environment, it sometimes appears that we are doing our very best to consign human beings to the dustbin of history along with the Dinosaurs. Our nature seems to be destructive, not just of our fellow man, but of the very planet we inhabit. Deep within us, like some long and half forgotten memory, mankind seems to have a desire to destroy not only the human race, but life itself. Why?

That mankind can be the so indifferent to the plight of his fellow humans and the rest of the life forms on this planet shows an unbelievable arrogance and greed. It may be, however, that in a delusory way we consider ourselves masters of all we survey and think it our rite to dominate, exploit and abuse everything around us for our own ends. This is criminal behaviour.

It's now obvious that most scientists on the planet are in agreement regarding the problems that the world faces and the speed that these changes will occur. Professor Timothy Lenton, of the University of East Anglia in England says :- "Climate change is likely to result in sudden and dramatic changes to some of the major geophysical elements of the Earth."

Professor James Lovelock, author of the Gaia Hypothesis and one of the world's top environmental scientists says :- "Climate change is now past the point of no return."

Peter Schwartz and Doug Randall in a report commissioned by the Pentagon say :- "Abrupt climate change could bring the planet to the edge of anarchy as countries defend and secure their dwindling food, water and energy supplies. The United States and Europe would become "virtual fortresses" to prevent millions of migrants from entering after being forced from land drowned by sea-level rise or no longer able to grow crops."

Sir Nicholas Stern said at the Royal Economic Society lecture at Manchester University, that :- "We risk damages on a scale larger than the two world wars of the last century."

Finally, James Hansen, the Director of NASA's Goddard Institute for Space Studies, (www.giss.nasa.gov) talking about the possibility of the Gulf Stream stopping, says :- "It would take no more than a quarter of 1 per cent more fresh water flowing into the North Atlantic from melting glaciers to bring the northwards flow of the Gulf Stream to a halt."

This would send northern Europe into a new Ice Age.

The question you have to ask yourself now is simple. "Do I believe what the scientists say?" If you don't, then I suppose you will be doing nothing, believing that the future will be bright and everything in the world will be cool. As Albert Einstein said :- "Great spirits have always encountered violent opposition from mediocre minds."

However, if you DO believe the scientists, then you have to make some of the biggest decisions in your life, because fairly shortly the shit is going to hit the fan. The other issue you have to face is a little more problematic. If you do believe what the scientists have to say, then you have to realise that mankind has now reached a point where only one thing seems absolutely certain. We are out of time and on the road to Hell.

Have a nice day now. ☺

Ian Gurney spent four years researching this book, contacting major government and United Nations organisations, major environmental scientists, climate scientists, doctors, politicians and a host of other scientific specialists on the problems of climate change, global warming, sea level rise, water shortages and world-wide food depreciation. Many thanks to all who helped in the research of this book.

THE END

www.ingramcontent.com/pod-product-compliance
Lightning Source LLC
Chambersburg PA
CBHW060550200326
41521CB00007B/545